Sonic Skills

Some Skills

Karin Bijsterveld

Sonic Skills

Listening for Knowledge in Science, Medicine and Engineering (1920s–Present)

Karin Bijsterveld
Faculty of Arts and Social Sciences
Maastricht University
Maastricht, The Netherlands

ISBN 978-1-349-95920-4   ISBN 978-1-137-59829-5   (eBook)
https://doi.org/10.1057/978-1-137-59829-5

This Palgrave Macmillan imprint is published by the registered company Springer Nature Limited
The registered company address is: The Campus, 4 Crinan Street, London, N1 9XW, United Kingdom

ACKNOWLEDGEMENTS

The research on which this essay draws, has been funded by the Netherlands Organisation for Scientific Research (NWO). In 2010, it awarded me with a generous VICI grant for the "Sonic Skills" project (277-45-003). The process of actually finalizing this essay has additionally been supported by the Netherlands Institute for Advanced Study (NIAS) through an individual fellowship in Amsterdam in the Spring of 2017.

The VICI grant enabled me to compose the Sonic Skills research team, and collaborate with its members Joeri Bruyninckx, Anna Harris, Stefan Krebs, Alexandra Supper and Melissa Van Drie. This essay could not have been written without their excellent research and publications—more on these scholars in the first chapter. Aline Reichow, student in the Maastricht research master program Cultures of Arts, Science and Technology, did exploratory research that found its way in Chapter 5. Jo Wachelder contributed by acting as a highly inspiring Ph.D. co-supervisor. The board and staff of the Faculty of Arts and Social Sciences at Maastricht University did everything in their power to support our activities.

One particularly memorable event was the Sonic Skills expert meeting held in Maastricht in January 2014. During this meeting, an international group of scholars commented on our project papers, for which I am deeply grateful. These scholars were Lisa Gitelman, Peter Heering, Aleks Kolkowski, Julia Kursell, Cyrus Mody, Rachel Mundy, Trevor Pinch, Tom Rice, Sophia Roosth, Jonathan Sterne, Axel Volmar, and our Maastricht colleagues Wiebe Bijker, Jens Lachmund, and Jessica

Mesman. I also genuinely enjoyed our 2015 Sonic Science Festival and Exhibition, organized with Marith Dieker's invaluable assistance.

Chapter 3 is largely based on the following open access article: Supper, Alexandra & Bijsterveld, Karin (2015). Sounds Convincing: Modes of Listening and Sonic Skills in Knowledge Making, *Interdisciplinary Science Reviews*, 40, 2, 124–144. I would like to thank Alexandra Supper for allowing me to reuse this article.

Finally, I am indebted to Rein de Wilde and, once again, Joeri Bruyninckx for their insightful comments on this essay, to Kate Sturge for her outstanding editing work, and to Palgrave editor Holly Tyler, editorial assistant Joanna O'Neill and editor Joshua Pitt for their guidance in making the manuscript into a book.

CONTENTS

CHAPTER 1

Listening for Knowledge: Introduction

Abstract We tend to associate the sciences with seeing—but scientists, engineers, and physicians also use their ears as a means for acquiring knowledge. This chapter introduces this essay's key questions about the role of sound and practices of listening in the sciences, and explicates their relevance for understanding the dynamics of science more generally. It defines the notion of sonic skills, situating it in the wider literature on the auditory dimensions of making knowledge. It presents the case studies on which the essay draws, explaining their geographical, temporal, and methodological scope and the researchers behind them.

Keywords Listening for knowledge · Sonic skills · Science · Medicine · Engineering

ACOUSTIC SIGNATURES

On the afternoon of July 11, 2014, Dutch Public Radio 1 broadcast an interview with science journalist Diederik Jekel. He had breaking news: American geologists had discovered a "super ocean" some 300 miles below the earth's surface. The journalist immediately added a qualification. What the Americans had actually found were some stone minerals, originating from the earth's deep layers, that included water molecules. This prompted the talk show host to ask how certain scientists could be of the super ocean's existence. The journalist explained that the

American geologists knew they had *in fact* found water when they sent sound waves deep into the earth. The stones had melted in the earth's heat, and just as a knock on a table sounds different from a knock on a glass of water under the same conditions, the melted material sounded different from the non-melted. The talk show host was quick to conclude: "We know it," she said, "but we have not seen it; it has not been [proven] experimentally."[1]

Apparently, she had trouble believing the geologists' ears. Their findings had not yet been *proven*, because the phenomenon had not been *seen*. By suggesting that hearing something is not sufficient to prove its existence, whereas seeing it would actually establish the fact, the interviewer posited a direct link between seeing and true science or ultimate knowledge. She may have learned to do so from scientists themselves, who tend to work in offices packed with printouts and scans around computer screens, producing publications rich in diagrams, graphs, and other images. Indeed, the American geologists in search of water had used seismic data gathered during earthquakes, often referring to infrasound waves: sound waves below the human audible range. At times, these infrasound waves are translated into frequencies that humans can hear, but more often they are made *visible*, in graphs. In the medical world, ultrasound waves—sounds above the human audible range—are used to create images of the body's interior, such as images of the unborn. Even scientists interested in frequencies that are directly audible to them, like most of the sounds of birds or language, usually turn their sound recordings into images before they start analyzing the objects of their interest. Some such images, including spectrograms and sonograms, also make their way into the scientists' publications and presentations, often accompanied by other forms of visual data representation.[2]

Nevertheless, scientists do listen for knowledge. In early 2015, an international group of geophysicists published an article claiming that particular patterns in the sounds of glaciers might reveal where and how those glaciers were calving. They had made sound recordings with hydrophones—underwater microphones—and taken photos at the same time. This enabled them to link various glacier sounds to distinct forms of ablation through "acoustic signatures" that indicated, for example, whether the ice was disappearing below or above the water surface (Glowacki et al. 2015). Other physicists monitor the condition of dikes by recording the structures' inner sounds with microphones, while biologists listen to the sounds of whales, insects, and birds.[3] In military contexts, too, scientists of various backgrounds have left their mark on listening technologies. During World War I, detecting and locating hostile submarines, artillery, tunnel-building, zeppelins, and bomber aircraft involved not only listening in on the enemy's wireless communication, but also developing hydrophones, geophones, sound-ranging equipment, and sound locators, as well as training personnel for such mediated listening—listening, that is, enhanced by acoustic and electroacoustic means. In the interwar years, huge acoustic mirrors were built to detect aircraft, while sonar (in full, Sound Navigation and Ranging) was designed to locate enemy vessels in both world wars and beyond.[4] Most of these technologies were "passive," in the sense that they merely captured the sounds produced by the objects of their interest. But sonar can also be "active," generating sound to detect objects through echolocation. In both cases, listening was a key dimension of the military use of these technologies.

Listening for knowledge is also embedded in more everyday practices. Since the early nineteenth century, doctors have used stethoscopes to listen to their patients' hearts and lungs as a way of investigating their health. Engineers in the automotive industry and mechanics in garages use automotive stethoscopes to listen to the functioning of car engines or the car's other moving parts. Another well-known tool is the Geiger counter, a device that transforms data about radiation not only into numbers but also into clicking sounds, informing and warning us of what we cannot see or sense in other ways. This capacity makes the Geiger counter, or more correctly the Geiger-Müller counter (Volmar 2015), a device for "sonification": "the use of nonspeech audio to convey information" (Kramer 1999). And yet when today's general practitioners or hospital specialists listen to our body and hear something seriously wrong, they will probably suggest checking our blood with the help of chemical analysis or "looking inside" with help of X-ray technology, magnetic resonance imaging (MRI), or other devices resulting in images. Even the glacier researchers emphasized that their tests would have to be repeated before anything "definitive" could be said. Capturing sound is one thing; interpreting and assessing the results is quite another, and is highly dependent on the contexts in which listening takes place.

This book-length essay aims to understand the ambiguous and at times contested position of listening for knowledge in the sciences. It does so by tracking the shifting status of sonic skills in science, medicine, and engineering across the long twentieth century, the theme of a project at Maastricht University that I will describe in more detail below. The project was primarily interested in sound and listening as a way of acquiring knowledge about human bodies, animals, machines, or other research objects, and thus in sound and listening as a *means* rather than an *object* of research. For our project's selection of case studies, this meant leaving out fields that prioritize the understanding of sound, hearing and listening per se, such as acoustics, psychoacoustics, otology, and audiology. Of course, theories and techniques from these fields have affected listening for knowledge in other areas (Kursell 2008; Hui 2013; Hui et al. 2013; Erlmann 2010). But the project's main focus is not research *about* sound, hearing and listening. Instead, *practices of listening in* the sciences take center stage—specifically, listening practices applicable to sounds in the frequency ranges audible to humans. And rather than dealing with the forms of listening that people engage in when interacting in conversations, this study is about listening, within the sciences, to the sounds of phenomena that do not talk back. The special relevance of listening's fluctuating standing in the sciences, this essay claims, is that it offers new insights into the significance of timing, trust, and accountability in knowledge making.

Sonic Skills

This essay, then, aims to study listening for knowledge in the sciences by focusing on the use of "sonic skills." Sonic skills, as I use the term here, include not only listening skills, but also the techniques that doctors, engineers, and scientists need for what they consider an effective use of their listening and recording equipment. Examples of such skills would be the proper positioning of a stethoscope on a patient's body, the handling of magnetic tape recorders in bird sound recording, or simply archiving sound samples for easy retrieval. To understand listening for knowledge, therefore, we need to study not only the skills related to listening proper, but also those that ensure sounds can be amplified, captured, reproduced, edited, compiled, accessed, and analyzed.

These skills can be examined as embodied and encultured techniques, just as Jonathan Sterne did in *The Audible Past* (2003), where he introduced the notion of the "audile technique" to articulate his interest in the context-specific bodily postures and usages of mediating instruments that are intended to isolate, intensify, and direct acts of listening. I apply a similar analysis to the sonic skills of making, recording, storing, and retrieving sound *in addition to* listening to sound. This implies that listening cannot be studied in isolation from the other senses. Its historically embedded relations with other sensory modalities, especially tactile and visual modalities, are crucial (see also Krebs 2015; Supper 2016).

Three clusters of issues are of particular interest here. First, for what purposes have scientists, engineers, and physicians lent their ears to the objects of their interest? Did that listening elicit new types of questions, and if so, of what kind? Second, how exactly have these experts employed their ears to make sense of what they studied? In what ways, and with the help of what tools, did they listen to what they examined? And how did—and do—they master such sonic skills? Third, under what conditions have sonic skills, alongside visual ones, been accepted as "objective" paths of inquiry in science, engineering, and medicine? And under what conditions did visual skills partially displace sonic skills again?[5] Why, for instance, did the use of sound spectrograms become widespread in ornithologists' research practices after the 1950s, whereas just a few decades earlier, in the late 1920s, bird researchers had welcomed film sound recorders and electrified gramophones with great enthusiasm? And why, at least at first glance, has the auditory presentation of scientific data still not become very popular, apart from warning devices such as the Geiger counter?

Although I will address these questions in relation to the long twentieth century—that is, including the late nineteenth century and early twenty-first century—I concentrate on the period beginning in the late 1920s. This was when the use of sound recording technologies such as the electric phonograph and gramophone took off in many areas of research, including ethnology, ethnomusicology, and ornithology, affecting both the listening practices and scholarly debates in these fields (Stangl 2000; Sterne 2003; Mahrenholz 2008; Mundy 2009, 2010). In the early days of ornithology, for instance, making field recordings of birds still required a seven-ton truck carrying a disc-cutter, microphones, cables, and an electric oven for softening wax. As a result, only the most urbanized parts of the wild could be studied—which of course affected the content of the recordings. The rise of the magnetic tape recorder in the early 1950s reshaped listening practices in ornithology once again, enabling collaboration with amateurs, storage of recordings in large sound archives, and the comparison of recordings originating from a wide variety of places. And from the late 1950s, ornithologists could turn to the sound spectrograph, a device that transformed audio recordings into sound spectrograms, triggering new standards for valid scientific proof (Bruyninckx 2013).

In this book, I will often refer to "the sciences" as a shorthand for science, medicine, and engineering. This does not imply, however, that the dynamics of auditory epistemology in the science-based professions, such as medicine and engineering, has been the same as in the natural sciences themselves. Despite fears that medical auscultation might be "a dying art" (a discourse recently analyzed by anthropologist Tom Rice 2013: 164), learning to listen with the stethoscope is still a standard component of medical curriculums worldwide. Similarly, those working in car repair shops and the automotive industry consider the mechanic's stethoscope an essential tool,[6] the use of which is an acknowledged learning-by-doing aspect of practical training (Krebs 2012). In contrast, as far as I am aware there are no courses in listening, or acknowledgement of the need for listening skills, in the science programs at universities. This distinction is just one obvious difference between the natural science professions—fields with legally assigned jurisdiction concerning particular forms of expertise (Abbott 1988)—and the sciences themselves.

As I hinted above, this essay is the outcome of work by a group of researchers at Maastricht University who for several years participated in the Sonic Skills project.[7] They include historian of science Joeri Bruyninckx, medical anthropologist Anna Harris, historian of technology Stefan Krebs, sociologist Alexandra Supper, and musicologist Melissa Van Drie. In 2008–2009, I outlined the project's questions, key concepts, case studies, methods, and theoretical approaches in a grant proposal for which I was awarded a VICI grant by the Netherlands Organization for Scientific Research (NWO) in 2010.[8] Joeri Bruyninckx and Alexandra Supper were both PhD students at the start of the project, and also contributed as postdocs in later phases; the others were involved as postdoctoral researchers. I coordinated the project from its inception until its completion in mid-2015.

This essay aims to synthesize the insights from the project's case studies. It draws on the two Ph.D. dissertations and the journal articles, book chapters, and outreach activities that resulted from the project, as well as the rapidly expanding body of literature on the epistemology of sound and listening written by colleagues elsewhere. In essence, my way of referring to the literature produced within the Sonic Skills project does not differ from my use of secondary literature produced outside of the project. But the Sonic Skills publications have defined the scope of my arguments, which are grounded in a systematic comparison of the project cases. Moreover, my role as principal investigator has given me privileged access to many of the interviews, observations, and historical sources behind the publications, leading to a deeper understanding of the project outputs than is possible for external publications. Occasionally, I will refer to these original sources and data with quotations that are more extensive than the ones to be found in the project publications.

To develop my argument here, I also rely on an article about listening modes in the sciences that I co-authored with Alexandra Supper and was published in *Interdisciplinary Science Reviews* in 2015. The key sections of that article constitute the heart of Chapter 3, now complemented by a discussion of the incidence of particular modes and some additional examples of listening. I am grateful to Alexandra for allowing me to reuse this article. I owe a great deal to the other Sonic Skills researchers as well. We devoted several meetings to the issue of listening modes by exchanging examples and refining the initial definitions while making use of the significant adjustments to these notions in Supper's dissertation. I also benefited tremendously from the Sonic Skills expert meeting mentioned in the acknowledgments.

By comparing the cases and claims from the earlier publications, I intend to answer our project's overarching questions about the use and epistemology of sonic skills in the sciences. Ideally, however, this essay will also lead readers to, or back to, the rich original work that underlies it—work that offers much more than the insights brought together here. Below, I will refer to the "we" of the project participants several times, for instance in Chapter 3. But let me first explain our historiographical and theoretical points of departure, and how these are rooted in previous scholarship on science, medicine, and engineering.

SENSORY PRACTICES IN THE SCIENCES

It is hard to find a scholar today who is willing to defend the claim that the dominance of visual forms of representation and proof in the sciences is best explained by the essential qualities of seeing, hearing, or other forms of sensory perception. Or, to be more precise, it is hard to find such a person in the academic fields that most strongly inform this essay: history of science, technology, and medicine; the interdisciplinary field of science and technology studies; history and anthropology of the senses; and constructivist strands within sound studies.

In cultural studies, art theory, and acoustic ecology, the situation is different. In those areas, there are indeed authors who argue that seeing creates the kind of distance from objects that scientists are after, whereas hearing enables a more intimate relation with the world. It was the pioneer of soundscape research, composer and acoustic ecologist Raymond Murray Schafer, who—inspired by Marshall McLuhan's work—distinguished between an "inward" drawing ear and an "outward" looking eye (Schafer 1967: 2; 1994/1977). Because of this orientation, he treated hearing as the *better* sense, and so did many of his followers. In this type of storyline, dubbed "the audiovisual litany" by Jonathan Sterne, seeing creates distance, calls upon the intellect, and focuses on the superficial, whereas hearing surrounds us with sounds, is inclined to the affective, and penetrates deep into the heart of the matter (Sterne 2003: 16). More recent varieties of this way of thinking are less schematic, but still underline the apparent natural affinity of the auditory with associative thinking and unfixed positions (Labelle 2011: xviii–xix), or present the auditory as troubling "the visually inspired epistemologies that we take for granted: the clear distinction between subject and object, inside and the outside, self and the world" (Bull 2006: 112). Immersion in sound, declares the "new orthodoxy" in sound art as critically discussed by music scholar Will Schrimshaw (2015: 155), is an experience that precludes reason.

In the fields with which our project is affiliated, however, most scholars seek *contextual* explanations for the significance of visual strategies in academic knowledge making. In the history of science, for instance, three interacting shifts have been held responsible for the dominance of visualization. In the seventeenth century, natural philosophy's focus on hermeneutic readings of the world and natural history's interest in classification gradually gave way to a new culture of science in which experiment and observation became important requirements for legitimate knowledge (Pickstone 2000, 2007). Who witnessed what, and where, made all the difference to an experiment's scholarly validity. Early reports of experiments in the meeting rooms of academic societies, published as letters, recounted which learned men had actually attended and added detailed images of the events in order to enable "virtual witnessing" (Shapin and Schaffer 1985: 60). This focus on eye-witnessing did not mean the reports neglected the sounds of the experiments (Schwartz 2011: 93–95). But over time, scientists would increasingly describe their work in visual terms.

Historians of science regard two other developments as relevant for understanding this development. The first is the rise of print in the fifteenth century and the growing availability of books thereafter.[9] The enhanced circulation of printed materials enabled texts, calculations, and illustrations to be systematically compared, triggering critical reflection and expanding the academic community. As time went by, the easy exchange of printed reports also naturalized a new notion of "witnessing," one that involved reading proceedings as opposed to attending the experiments proper (Johns 1998). The second relevant change is the emergence of "mechanical objectivity" as a epistemic ideal in eighteenth and nineteenth century science. This dismissed the human body's status as a trustworthy witness of natural phenomena, favoring instead their automatic registration by machines. Mechanical objectivity, Lorraine Daston and Peter Galison (1992, 2007) have shown, experienced its heyday in the 1920s, after which "trained judgment" by expert scientists gradually acquired validity—at least as a supplement to mechanical objectivity. Still, for a long time, doing science was largely about reading the instruments that registered natural phenomena for the scientists—a visual activity indeed. Furthermore, many of these instruments translated what they registered into graphs, and thus visual representations.

By the time these studies in the history of science appeared, science and technology studies (STS) had entered the stage, the product of a genuine interest in the everyday practices of science. Until the 1980s, the study of science had striven to demarcate rational science from irrational beliefs by formulating universal criteria for scientific knowledge. Scholars in STS moved away from over-idealized conceptions of science and turned their attention to "science-in-action," a term coined by Bruno Latour. Latour and his colleagues aimed to trace how science was done in practice, in the laboratory and beyond. One of their questions was how scientists managed to turn local findings into global truths. Latour's own answer contributed importantly to understanding the role of visualization in scientific representation. He showed that inscriptions, such as tables and diagrams, can effectively circulate locally acquired data across geographically disparate networks of knowledge because they are both immutable and mobile. And he pointed out that such inscriptions can be easily cascaded and superimposed on paper—an example being numeric data printed on maps (Latour 1986, 1987).

It is not that the interpretation of these visual resources was, or is, always immediately self-evident to the community of scientists involved.

After any presentation at any science conference, many questions will focus on the proper reading of the graphs. Similarly, the introduction of visualization techniques such as photography, X-ray, probe microscopy, or planet observation by panoramic cameras has never come with instant transparency or an obvious way to decode the resulting images. On the contrary, each new visualization strategy gives rise to fierce debate as to what exactly the images represent. Novel images have to be made commensurable with existing representation techniques in order to become legible (Pasveer 1989, 2006; Te Hennepe 2007; Mody 2014), and that entails new skills: formerly unknown features of a phenomenon only "pop out" through digital image processing, for example, because "a skilled vision is crafted into the image from the outset" (Vertesi 2014: 25).

Research into visualization techniques has thus significantly refined our understanding of how knowledge and visual displays interact. Michael Lynch (1990), for instance, explained that interaction by describing how diagrams and images in the life sciences act as an "externalized retina" that structures the production of scientific facts through the schematic redefinition, mathematical order, and solidification of the objects under study. If Lynch focused on the *conventions* of "consensual 'seeing' and 'knowing'" in the sciences (p. 155), Eugene Ferguson studied nonverbal "visual thinking" in design at the individual level (Ferguson 1992). Such visual literacy did not remain limited to scientists, but gradually spread among the general population through the consumption of visual toys and other forms of educational entertainment (Stafford 1994; Tufte 1997).

STS scholars' work on the laboratory practices of scientists unveiled much more than the visual aspects of knowledge making, however. As historian of science Lissa Roberts (1995) suggested, it was only in their published writings that eighteenth-century chemists sidelined touch, hearing, smell, and taste. While carrying out their experimental work, they still enlisted their senses to interpret what had happened, carefully attuning their bodies to their instruments. Harry Collins's field studies of the lab work of present-day physicists showed how the "tacit knowledge" that scientists develop about their experimental set-ups often makes it impossible for others to fully replicate their experiments. The result is an infinite "experimenters' regress" (Collins 1985). Collins drew his notion of tacit knowledge from the work of Michael Polanyi (1983/1966), and initially defined it as "an embodied kind of know-how irreducible to symbolic terms" (Collins cited in Mody 2005: 176; Collins 2001).

It is wise to note at this point that not all science draws on embodied knowledge. Think of the cognitive focus in contemporary mathematics—even if much mathematics-in-action does involve enacting arguments through jottings, annotations, and erasures on blackboards and scrap paper (Barany and MacKenzie 2014). But despite exceptions such as most of mathematics, the discussion of embodied and tacit knowledge has been an important source of inspiration to STS scholars researching the role of the senses other than seeing in science, medicine, and engineering. Natasha Myers (2008), for instance, has remarked on the significance of *gestural* knowledge in the crystallography of proteins. Crystallographers use graphical software to model their proteins in three dimensions, but the most experienced among them check the draft models by imagining the protein dimensions and characteristics in terms of their own corporeal experience of building earlier protein models. They also use their bodies as resources for communicating a "feel" for molecular structures to novices.[10]

Similarly intriguing is Sophia Roosth's (2009) account of "sonocytology" in nanotechnology research, whereby cell wall vibrations are recorded with scanning probe microscopes and amplified to volumes audible to humans. In the first years of the twenty-first century, she explains, the US scientist Jim Gimzewski introduced sonocytology as a noninvasive technique for studying cellular interiors, contrasting with invasive forms of research such as chemical analysis. He initially studied yeast cells, but later began to listen diagnostically to the "difference between healthy and cancerous cells" (p. 341). Gimzewski interpreted the sounds as referring to particular forms of cellular motion and metabolism. The epistemological effect of studying cells in terms of sound, Roosth argues, was to conceive of cells "in time and in context," or, more specifically, in terms of interior time and—with an ear for the transmission of sound—acoustic micro-milieus (p. 338). And when Gimzewski claimed to hear his yeast cells "screaming," he also anthropomorphized them (p. 339).

This work is of fairly recent date, but the theme of listening in science, engineering, and medicine entered the STS research agenda much earlier, often linked to an interest in tacit knowledge. Sociologist of science Jens Lachmund (1994), for example, studied the rise of the stethoscope and auditory knowledge in nineteenth-century medicine by examining the work of the Parisian physician René Laennec and his Viennese colleague Joseph Skoda. Laennec not only invented the stethoscope in 1816, but also created the very first codification of lung sounds—a codification that was intended to help physicians relate the body's murmurs to its medical condition. In an attempt to explain the nature of these sounds to uninitiated physicians, Laennec compared them with the sounds of animals, musical instruments, and urban life, and occasionally used musical notation. In his view, however, such codifications could never be sufficient to learn auscultation: hospital training was indispensable for acquiring the right skills (Lachmund 1994, 1999).

Jacalyn Duffin (1998) finds that attention to Laennec's musical skills is crucial to understanding his successful use and teaching of the stethoscope. However, those skills were not enough to disseminate Laennec's lung sound codification beyond the Paris hospitals. The hospital and research practices of Skoda's Vienna, for instance, were rather different from those in Paris, with limited access to patients but more extensive connections with laboratory research. These divergences gave rise to different types of lung sound codification. Whereas Laennec had worked inductively and assumed a one-to-one relationship between sounds and pathologies, Skoda reasoned more deductively, developing an auditory form of differential diagnosis that started by excluding potential causes with the help of acoustic knowledge of the body and its inner resonances (Lachmund 1994, 1999). Although I will return to Lachmund's research in the next chapter, it is worth discussing in some detail here, for it neatly articulates the way that the rise of a new instrument afforded new listening practices and forms of auditory knowledge, but did not simply determine the *character* of those practices and knowledge. In different settings, the same instrument may lead to diverging epistemologies.

The stethoscope happens to be one of those auditory instruments that turned up in various contexts, and not only medical ones. As early as the 1920s, technical literature on car manufacturing told of engineers listening to car engines in order to detect problems in the machines. To track the car's lowest frequencies, for instance, they used a metal rod that transmitted the engine's vibrations to the engineer's teeth (Snook 1925). Over time, car manufacturing and repair businesses also started to work with automotive stethoscopes, with the deliberate intention of transferring the prestige of the stethoscope, the "hallmark of a doctor" (Rice 2013: 74), to the profession of the mechanic (Krebs and Van Drie 2014). But it was more than just an icon of expertise: car mechanics listened to engines to detect the causes of flaws in the first place (Borg 2007).

Listening to machines was not limited to the automotive industry. Gerard Alberts, a historian of information technology, has explained how 1950s operators of digital computers at the Philips Physics Laboratory in the Netherlands missed the "trustworthy" rattling sounds of electro-mechanical calculators and decided to make the computers' calculation processes audible through speakers. Adding loudspeakers to the computers created an "auditory monitor," restoring a sensory relation with the equipment and making it possible to listen to computers in order to debug them (Alberts 2000, 2003: 17, 23). Listening has also been mentioned in the wake of scholarly interest in "situated actions" in engineering and repair (Suchman 1987). For technicians servicing photocopiers, Julian Orr has shown, "the succession of noises narrates to the experienced ear the progress of the operation" (Orr 1996: 98; Pacey 1999). This is why the technicians Orr studied hated noisy customer sites: the noise hampered their auditory focus on the machines. For similar reasons, factory workers long resisted the use of ear plugs, which deprived them of auditory cues about how well the machines on the shop floor were working (Bijsterveld 2008, 2012).

Patterns of sound are no less relevant at the laboratory bench. Investigating materials science, Cyrus Mody has discussed how the sounds of the valves, pumps, and outputs of laboratory instruments give the staff information on the experiments' quality and content: "Learning these sounds and the experimental rhythm they indicate is part of learning the proper use of the instrument," including "tacit knowledge of the sounds made when tools are not operating smoothly" (Mody 2005: 186). Some laboratory employees say that data expressing periodicity are much better processed with the ears than with the eyes. Intriguingly, they also find that the aesthetics of sound enhances "embodied interaction with the instrument" (p. 188).

The growing interest of STS scholars in the everyday practices of science, medicine, and engineering has thus both articulated the relevance of the nonvisual senses for knowledge making and explained why that relevance does not become immediately apparent when reading scientists' work or attending their presentations—many listening activities take place behind the scenes of science, but are no less important for that science. In the history and anthropology of the senses, there has been a similar acknowledgment of the multimodality of sensory orientation. For many years, historians and anthropologists seemed almost obsessed with characterizing particular periods or even entire cultures in terms of a predominating sense. Cultural historian Peter Bailey, for instance, distinguished between the modern West, which he believed to have had a visual focus ever since the advent of print, and "pre-modern societies," which had been "predomi-nantly phono-cen-tric, privileging sound over the other senses in a world of mostly oral-aural communi-cati-on" (Bailey 1996: 55). Anthropologists, in contrast, claimed that sight had held the highest position in Western cultural representations of the senses since antiquity (Classen 1997), or discussed present-day cultures with alternative sensory orientations, such as the Kaluli people in the tropical forests in Papua New Guinea and their auditory epistemology—or "acoustemology" (Feld 2003; Feld and Brenneis 2004).

Today, however, it seems that more and more scholars are adopting Jonathan Sterne's critique of the assumption that the history of the senses should be "a zero-sum game, where the dominance of one sense by necessity leads to the decline of another" (2003: 16). Some of these scholars have pointed out the significance of sound for Westerners' everyday spatial and symbolic orientation in the nineteenth or twentieth century (Corbin 1999; Bull and Back 2003); others argue that a "perceptual equilibrium" has been present since at least the later medieval period (Woolf 2004). Anthropologist Tim Ingold has taken this one step further: instead of studying cultures as mere filters of sensory experience, we should examine how people are informed by their senses, and by all their senses together, as they are *moving* through particular worlds or cultures—worlds that themselves have particular materialities. This renders the idea that seeing is a static and distancing experience unconvincing: it unjustly conflates *seeing* with *visualization*. We constantly move our head or focus in and out when looking at something; only our drawings and pictures solidify a particular perspective. Moreover, studying one sensory modality, such as hearing, makes no sense if we fail to acknowledge its interconnections with other sensory modalities (Ingold 2000, 2011a, b).

These remarks bring me full circle, back to the predominance of anti-essentialist approaches with which I started this section. Those assumptions were the starting point of the Sonic Skills project as well. We insisted on a practice-oriented approach that keeps an eye and an ear open for the role of tacit knowledge in the sciences and for scientists' use of their senses as they move around in different settings. Although several useful studies on listening in the sciences have been published, they leave many questions open. What about the commonalities between various strategies of listening, for instance? The role of music? The differences between professional contexts and the academic sciences? I will return to these systematic issues below. But let me first discuss our project's selection of the case studies.

CASES OF SOUND AND LISTENING

To attain a better understanding of the role of the senses in knowledge dynamics, we considered it important to select a variety of sites where scientists, engineers, and physicians perform their work. Among other parameters, those sites vary in terms of public accessibility—the literature suggests a difference between the public presentation of science and what happens behind the scenes. This inspired us to choose both settings where experts are among colleagues (such as factories, laboratories, and the field), and more open environments, such as hospitals, where lay people are present, or conference halls and other venues at which scientists present their results to the wider world. This enabled us to include contexts of both professional expertise (hospital, factory) and scientific expertise (field, lab, conference). The multisite approach also allowed us to cover different phases in research and design, ranging from tinkering with technology on shop floors to trial-and-error in laboratories, from recording natural phenomena to displaying scientific data through sonification.

Our preference for particular cases within each of these sites resulted from several other considerations. One was the idea of encompassing a wide array of sonic tools: technologies that have allowed scientists, engineers, and physicians to focus on, amplify, record, or transform sound, such as listening rods in engineering, stethoscopes in medicine, tape recorders in field research, or software in sonification. Another was the wish to understand how novices *acquire* sonic skills, most notably in hospital settings. With Brian Kane (2015: 8), we believe that in the process of cultivating such skills, "much of the cognitive effort involved in the initial training is offloaded onto the body." Observing the training of sonic skills before they are naturalized opens them up for analysis. This objective is akin to Thomas Porcello's work on how studio engineers learn to understand sound (Porcello 2004) and to the work by Tom Rice and John Coltart on how medical students come to handle the stethoscope (Rice and Coltart 2006; Rice 2013). Whereas Rice and Coltart focused on the use of the stethoscope in cardiology for diagnosing heart diseases, we addressed respiratory medicine and lung diseases.

It was these considerations that largely defined our geographic scope, as we followed the clusters of scientists, engineers, and physicians chosen across Western Europe, the United States, and Australia. Our attention to local specificities responded to Michele Hilmes's 2005 comments on early sound studies work, in which she warned against creating "a seemingly transhistorical, transcultural essentialism that is actually predicated closely on an American model" (Hilmes 2005: 258).

Based on these various concerns, we decided to focus on the listening practices of the following groups. Stefan Krebs studied engineers and mechanics in the automotive industry of Germany and France (1920s–present) and in a paper factory in the United Kingdom (1970s–present). Joeri Bruyninckx investigated ornithologists in field settings in the United States, the United Kingdom, and Germany (1920s–present), and present-day scientists in material physics and molecular biology in laboratories in the Netherlands and the United States. Melissa Van Drie asked how physicians listened, and taught medical students to listen, to their patients' lungs in France, the United Kingdom, and the United States (1950–2010), while Anna Harris did the same for contemporary doctors and students in the Netherlands (Maastricht University Skills Lab) and Australia (Royal Melbourne Hospital Medical School). Alexandra Supper examined the listening practices of sonification experts in Western Europe and the United States who were participating in the International Community for Auditory Display (ICAD), established in 1992. Finally, I researched cross-case issues such as listening modes, the verbal expression of sound, and notation, extending the range of examples while gathering historical information on shifts in sonic skills.

For nearly all cases, we combined traditional historical methods with ethnographic observation, interviewing, and reenactment. The historical methods included archival work, oral history interviewing, and analyzing published historical sources such as scientific and trade journals, textbooks, instruction manuals, and the published memoirs of ornithologists and recordists. Our ethnographic approaches included observing scientists, engineers, mechanics, medical staff, students and sonification experts in labs, on shop floors, in hospitals, and in presentation and performance venues; studying their logbooks, making recordings, taking pictures, and carrying out in-depth, semi-structured, and qualitative interviews with them.

We also investigated how past scientists created sound recordings or employed sonic research tools by reenacting their use of historical instruments ourselves or attending such reenactments. Work done by our colleagues Peter Heering and Aleks Kolkowski was a source of inspiration. Peter Heering is a former member of the "Oldenburg School" in the history of science, which analyzed historical experimental practices by replicating scientific experiments from the past, greatly improving historians' understanding of the affordances of particular tools and why they were worthwhile for scientists and wider audiences at the time (Heering 2008). Aleks Kolkowski is a researcher, artist, and violinist with extensive experience in reconstructing past practices of sound recording, using period technologies such as the mechanical phonograph, the electrical phonograph, and 78 rpm gramophones. He allowed Joeri Bruyninckx and other team members to witness him working with the tools of early ornithological recordists. In addition, Bruyninckx experienced the skills involved in using lab instruments by observing lab scientists and technicians working with those instruments. Stefan Krebs visited an old paper factory in the UK to listen to its machines together with the operators. Melissa Van Drie worked with the audio cassettes that had once instructed stethoscope-related skills to medical students, while Anna Harris co-listened to body and hospital sounds along with doctors, nurses, and students, at times employing the sounds she recorded on the wards as an elicitation technique in her interviews (Harris 2015). And Alexandra Supper learned to make sonifications in order to gain firsthand experience of the interrelated skills this process required. In all these situations, we were able to reinvoke some elements of the tacit knowledge associated with the activities, and to acquire better understanding of the distinctions applied by the people we studied.

An example of such deeper insights in the notions used in the communities examined is that one of our team members, Alexandra Supper, learned to distinguish between several sonification techniques. One is "audification," or "scaling existing vibratory signals into human hearing range" (Harris 2012: note 3), as when seismographic vibrations are transformed into audible signals in order to understand and predict the dynamics of earthquakes (Dombois 2001). Another is parameter mapping, whereby sound parameters such as pitch, duration, loudness, and timbre are controlled by the characteristics of the underlying data. Proponents of sonification claim that the auditory display of data is especially useful for an exploratory analysis of large, multivariate datasets, where certain patterns, such as variations on a time or spatial series, may be easier to detect by ear than by eye (Baier et al. 2007; Dayé and de Campo 2006). Despite such claims, sonification is still highly contested in the sciences, and is often treated as a form of music or sound art rather than as science proper. Indeed, techniques of sonification have also been employed by composers, occasionally in collaboration with scientists. In this sense, sonification is something of a "breaching experiment" (Garfinkel 1967), challenging taken-for-granted conventions in the sciences.

SENSORY SELECTIVITY

This essay's ultimate aim is to present insights into the issue of sensory selectivity—the high value attached to specific sensory modalities or their combinations—in the production and validation of scientific and professional knowledge. Rather than proclaiming a victory of the visual in science, pleading for the emancipation of hearing at the expense of seeing, or defending a perceptual equilibrium, I will investigate when, how, and under what conditions the ear has contributed to knowledge dynamics, whether in tandem with or as an alternative to the eye. To this end, I combine a synchronic analysis of listening modes and sonic skills with a diachronic analysis of how listening practices in the sciences have developed over time. The conceptualization of listening modes and sonic skills will help to refine my historical analysis and to systematize my comparison of cases, culminating in a theory that explains the shifting relevance and legitimacy of listening practices in science, technology, and medicine.

I devote an entire chapter to the notion and relevance of modes of listening, but would already like to mention here that our project distinguished between three purposes of listening—monitory, diagnostic, and exploratory listening—and three ways of listening: analytic, synthetic, and interactive. This novel classification of listening modes in science, engineering, and medicine has been derived both from primary sources and from the still scattered secondary literature on listening in the sciences and other domains, such as radio (Douglas 1999). Some of the modes of listening, as well as the ability to switch between different modes, are enabled by particular tools and by the sonic skills that scientists, physicians, and engineers have developed to handle them. I have already defined sonic skills as the skills required for making, recording, storing, retrieving, and listening to sound. They include the skills used by scientists for representing and sharing sound, such as the skills of recording sound with help of musical notation—as early ornithologists did.

These concepts of listening modes and sonic skills helped us to systematize our comparative analysis, informing a theory of sonic skills that identifies the conditions under which particular listening modes and sonic skills, as *ensembles* of listening for knowledge, have been accepted or rejected as a legitimate entrance to knowledge in the sciences. Our theory gives special prominence to three conditions. The first is the *timing* of interventions in science, technology and medicine, and how this has been afforded by tools such as stethoscopes, recording devices, and software programs. How did these relations between science and technology (in the form of tools) affect the status of listening in science, engineering, and medicine? Did the tools enable easy switching between modes of listening, afford a quick response to urgent issues, or alter the options for comparing data? The second condition is *trust* and the historically generated distinctions between the sciences and the professions. Work by Andrew Abbott (1988) on how professions claim and are endowed with irreducible, exclusive expertise and Richard Sennett's study (2008) on the significance of craft in professional work inspired us to ask why listening remained more significant in medicine and engineering than in some other contexts. In addition, Pierre Bourdieu's (1984) eye for the role of bodily discipline in acquiring a professional habitus, and the linkages between habitus and wider social practices, has helped us to understand the use of the stethoscope in both the world of doctors and the world of the engineers who tried to copy doctors. The third condition is the growing need for public *accountability* of science, and thus a shift in the relations between science and society. What does this imply for the position of music in the sciences, for instance? How did the putative links between sonic skills and musical abilities—such as the ability to notice and record differences in pitch, rhythm, or timbre—affect the acceptability of listening practices in science, technology, and medicine? Alexandra Hui has shown how in mid-nineteenth-century German and Austrian psychophysical research on the sensation of sound, musical skills were regarded as scientific skills. Yet by the end of that century, "the value of musical skill had become contested" among the researchers involved (Hui 2013: 145). Did this devaluation of musical skills in knowledge making continue in the twentieth and twenty-first century? If so, what should we make of scientists' eagerness to reach out to wider audiences by bringing sound and music into the equation?

The answer to this last question may clarify why the science journalist cited at the outset of this chapter chose to refer to the *sounds* of the super ocean deep in the earth, even though the scientists who actually published on the topic had been *watching* graphs of vibrations. Exploring such issues will help me to build a theory that distinguishes not only between synchronic listening modes and sonic skills, but also between diachronically changing relationships of science and technology, science and professions, and science and society—one that explains the shifting legitimacy of listening for knowledge and the changing ensembles of sonic skills in the sciences. All this will be brought together in the final chapter. First, though, in Chapter 2, I analyze how scientists, physicians, and engineers employed, talked about, and transcribed sound—issues that also allow me to introduce most of our case studies in more detail. Chapter 3 proceeds with a discussion of listening modes in the sciences. The argument then moves from a synchronic to a more diachronic perspective, as Chapter 4 analyzes the shifting conditions that explain why listening for knowledge has so often been contested, and Chapter 5 asks why listening nevertheless kept returning in the sciences.

NOTES

1. Radio1, VARA, *De Nieuws BV*, July 11, 2014, available at http://www. denieuwsbv.nl/Singleview.12722.0.html?tx_ttnews%5Btt_news%5D= 121545&cHash=9450eb67a786024bc91133b11557b309 (last accessed February 12, 2015). The group of scientists was led by geologist Steven Jacobsen from Northwestern University, Evanston, Illinois. The mineral rock was ringwoodite, and the water had been located at a depth between 410 and 660 km below the earth's surface (see also Coghlan 2014; Schmandt et al. 2014).
2. Spectrograms, or sound spectrograms, are three-dimensional graphs representing sound across time. Time is displayed horizontally (on the x-axis), the spectrum of frequencies is presented vertically (on the y-axis), and the sound's intensity is expressed as shades of gray. Frequency and intensity refer to the acoustic properties of sound, whereas pitch and loudness refer to the perception of these properties by humans. In the early years of the sound spectrograph, the terms *spectrogram* and *sonogram* were almost interchangeable. At times, sonogram was used for specific spectrograms, such as speech sonograms or bird sonograms. Today, the term sonogram is most commonly used for medical ultrasound images.
3. For the use of acoustic technologies in the monitoring of dikes, see http://www.dijkmonitoring.nl/index.php/dijkmonitoring-keuzetool/

29-meettechnieken/34-geluidsmetingen (last accessed March 16, 2015). For an overview of bioacoustics, see http://www.bioacoustics.info (last accessed March 16, 2015).

4. Geophones were employed to listen to the sounds of underground activities. These instruments resembled stethoscopes, but had microphone membranes to record the sounds and cable connections to transmit them (Encke 2006: 120). On other military listening devices created in World War I, see Rawlinson (1923: 112 and 103–120), Hoffmann (1994: 268), Volmar (2012, 2014), and Bruton and Gooday (2016a). Elizabeth Bruton and Graeme Gooday (2016b) specifically discuss listening to submarines. For listening during land-based combat, see Lethen (2000) and Schirrmacher (2016); in tunnels and trenches, see Encke (2006); and for air defense, see Judkins (2016). On acoustic mirrors for air defense in the interwar years, see Scarth (1999), Van der Voort and Aarts (2009) and, again, Judkins (2016). On the history of active sonar, see Hackmann (1984).

5. The relevance of these questions has also been mentioned in the *Oxford Handbook of Sound Studies*, edited by Trevor Pinch and Karin Bijsterveld (2012: 11–12).

6. See, for instance, https://www.youtube.com/watch?v=9JDhEwMS_Us (last accessed March 16, 2015).

7. See the Sonic Skills Project Website, http://fasos-research.nl/sonic-skills/, and the virtual version of the Sonic Skills Exhibition, at http://exhibition.sonicskills.org/ (both last accessed March 30, 2017).

8. Bijsterveld, Karin (2009). Sonic Skills: Sound and Listening in the Development of Science, Technology and Medicine (1920 to now). Grant Proposal, Netherlands Organisation for Scientific Research. Short version available at http://www.nwo.nl/en/research-and-results/research-projects/95/2300157595.html (last accessed March 16, 2015).

9. For anthropologist Stephen A. Tyler, in contrast, the defining moment for the "hegemony of the visual as a means of knowing/thinking" in the West is not the rise of print, but the rise of literacy (Tyler 1984: 23). Writing itself is a kinetic action, but one that ultimately reduces speech to a thing seen, "freezing thought in visible form" (Tyler 1984: 33). That is why, in Tyler's argument, the diffusion of literacy has pushed aside the use of verbal metaphor for thinking and knowing ("I say to myself" for "I think") in favor of visual metaphor ("I see" for "I understand").

10. For other work on how gestures contribute to the articulation of images, see Katja Mayer (2011) on node-edge sociograms and Morana Alač (2014) on functional Magnetic Resonance Imaging (fMRI).

REFERENCES

Abbott, A. (1988). *The System of Professions: An Essay on the Division of Expert Labour*. Chicago: University of Chicago Press.

Alač, M. (2014). Digital Scientific Visuals as Fields for Interaction. In C. Coopmans, J. Vertesi, M. Lynch, & S. Woolgar (Eds.), *Representation in Scientific Practice Revisited* (pp. 61–87). Cambridge: MIT Press.

Alberts, G. (2000). Computergeluiden. In G. Alberts & R. van Dael (Eds.), *Informatica & Samenleving* (pp. 7–9). Nijmegen: Katholieke Universiteit Nijmegen.

Alberts, G. (2003). Een halve eeuw computers in Nederland. *Nieuwe Wiskrant, 22*(3), 17–23.

Baier, G., Hermann, T., & Stephani, U. (2007). Multi-channel Sonification of Human EEG. In *Proceedings of the 13th International Conference on Auditory Display, Montreal, Canada, June 26–29* (pp. 491–496).

Bailey, P. (1996). Breaking the Sound Barrier: A Historian Listens to Noise. *Body & Society, 2*(2), 49–66.

Barany, M. J., & MacKenzie, D. (2014). Chalk: Materials and Concepts in Mathematics Research. In C. Coopmans, J. Vertesi, M. Lynch, & S. Woolgar (Eds.), *Representation in Scientific Practice Revisited* (pp. 107–129). Cambridge: MIT Press.

Bijsterveld, K. (2008). *Mechanical Sound: Technology, Culture and Public Problems of Noise in the Twentieth Century*. Cambridge: MIT Press.

Bijsterveld, K. (2012). Listening to Machines: Industrial Noise, Hearing Loss and the Cultural Meaning of Sound. In J. Sterne (Ed.), *The Sound Studies Reader* (pp. 152–167). New York: Routledge.

Borg, K. (2007). *Auto Mechanics: Technology and Expertise in Twentieth-Century America*. Baltimore, MD: Johns Hopkins University Press.

Bourdieu, P. (1984). *Distinction: A Social Critique of the Judgement of Taste*. Cambridge, MA: Harvard University Press.

Bruton, E., & Gooday, G. (2016a). Listening in Combat: Surveillance Technologies Beyond the Visual in the First World War. *History and Technology, 32*(3), 213–226.

Bruton, E., & Gooday, G. (2016b). Listening in the Dark: Audio Surveillance, Communication Technologies, and the Submarine Threat During the First World War. *History and Technology, 32*(3), 245–268.

Bruyninckx, J. (2013). *Sound Science: Recording and Listening in the Biology of Bird Song, 1880–1980* (Ph.D. thesis, Maastricht University).

Bull, M. (2006). Auditory. In C. A. Jones (Ed.), *Sensorium: Embodied Experience, Technology, and Contemporary Art* (pp. 112–114). Cambridge: MIT Press.

Bull, M., & Back, L. (Eds.). (2003). *The Auditory Culture Reader*. Oxford: Berg.

Classen, C. (1997). Foundations for an Anthropology of the Senses. *International Social Science Journal*, 49(153), 401–412.

Coghlan, A. (2014, June 12). Massive "Ocean" Discovered Towards Earth's Core. *New Scientist*. Available at http://www.newscientist.com/article/dn25723-massive-ocean-discovered-towards-earths-core.html?cmpid=RSS|N-SNS|2012-GLOBAL|online-news#.VK_BGyx0z9L. Last accessed March 13, 2015.

Collins, H. M. (1985). *Changing Order: Replication and Induction in Scientific Practice*. London: Sage.

Collins, H. M. (2001). Tacit Knowledge, Trust, and the Q of Sapphire. *Social Studies of Science*, 31(1), 71–86.

Corbin, A. (1999). *Village Bells: Sound and Meaning in the Nineteenth-Century French Countryside*. London: Macmillan.

Daston, L., & Galison, P. (1992). The Image of Objectivity. *Representations*, 10(40), 81–128.

Daston, L., & Galison, P. (2007). *Objectivity*. New York: Zone Books.

Dayé, C., & de Campo, A. (2006). Sounds Sequential: Sonification in the Social Sciences. *Interdisciplinary Science Reviews*, 31(4), 349–364.

Dombois, F. (2001). Using Sonification in Planetary Seismology. In *Proceedings of the 7th International Conference on Auditory Display, Espoo, Finland, July 29–August 1* (pp. 227–230).

Douglas, S. J. (1999). *Listening in: Radio and the American Imagination, from Amos 'n' Andy and Edward R. Murrow to Wolfman Jack and Howard Stern*. New York: Times Books.

Duffin, J. (1998). *To See with a Better Eye: A Life of R.T.H. Laennec*. Princeton, NJ: Princeton University Press.

Encke, J. (2006). *Augenblicke der Gefahr: Der Krieg und die Sinne*. München: Wilhelm Fink Verlag.

Erlmann, V. (2010). *Reason and Resonance: A History of Modern Aurality*. New York, NY: Zone Books.

Feld, S. (2003). A Rainforest Acoustemology. In T. Reader (Ed.), *Michael Bull & Les Back* (pp. 223–239). Oxford: Berg.

Feld, S., & Brenneis, D. (2004). Doing Anthropology in Sound. *American Ethnologist*, 31(4), 461–474.

Ferguson, E. (1992). *Engineering and the Mind's Eye*. Cambridge: MIT Press.

Garfinkel, H. (1967). *Studies in Ethnomethodology*. Englewood Cliffs: Prentice Hall.

Glowacki, O., Deane, G., Moskalik, M., Blondel, P., Tegowski, J., & Blaszczyk, M. (2015). Underwater Acoustic Signatures of Glacier Calving. *Geophysical Research Letters*, 42(3), 804–812.

Hackmann, W. (1984). *Seek and Strike: Sonar, Anti-submarine Warfare and the Royal Navy, 1914–54*. London: Her Majesty's Stationery Office.

Harris, A. (2015). Eliciting Sound Memories. *The Public Historian, 37*(4), 14–31.

Harris, Y. (2012). Understanding Underwater: The Art and Science of Interpreting Whale Sounds. *Interference: A Journal of Audio Culture.* Available at: http://www.interferencejournal.org/understanding-underwater/. Last accessed August 18, 2017.

Heering, P. (2008). The Enligthened Microscope: Re-enactment and Analysis of Projections with Eighteenth-Century Solar Microscopes. *British Journal for the History of Science, 41*(3), 345–367.

Hilmes, M. (2005). Is There a Field Called Sound Culture Studies? and Does It Matter? *American Quarterly, 57*(1), 249–259.

Hoffmann, C. (1994). Wissenschaft und Militär: Das Berliner Psychologische Institut und der I. Weltkrieg. *Psychologie und Geschichte, 5*(3–4), 261–285.

Hui, A. (2013). *The Psychophysical Ear: Musical Experiments, Experimental Sounds, 1840–1910.* Cambridge: MIT Press.

Hui, A., Kursell, J., & Jackson, M. (Eds.). (2013). Music, Sound and the Laboratory from 1750–1980. *Osiris, 28*(1), 1–11.

Ingold, T. (2000). *The Perception of the Environment: Essays on Livelihood, Dwelling and Skill.* London: Routledge.

Ingold, T. (2011a). Worlds of Sense and Sensing the World: A Response to Sarah Pink and David Howes. *Social Anthropology, 19*(3), 313–317.

Ingold, T. (2011b). Reply to David Howes. *Social Anthropology, 19*(3), 323–327.

Johns, A. (1998). *The Nature of the Book: Print and Knowledge in the Making.* Chicago, IL: University of Chicago Press.

Judkins, P. (2016). Sound and Fury: Sound and Vision in Early U.K. Air Defence. *History and Technology, 32*(3), 227–244.

Kane, B. (2015). Sound Studies Without Auditory Culture: A Critique of the Ontological Turn. *Sound Studies: An Interdisciplinary Journal, 1*(1), 2–21.

Kramer, G. (Ed.). (1999). *Sonification Report: Status of the Field and Research Agenda, International Community for Auditory Display.* Available at http://icad.org/websiteV2.0/References/nsf.html. Last accessed August 18, 2017.

Krebs, S. (2012). Automobilgeräusche als Information: Über das geschulte Ohr des Kfz-Mechanikers. In A. Schoon & A. Volmar (Eds.), *Das geschulte Ohr: Eine Kulturgeschichte der Sonifikation* (pp. 95–110). Bielefeld: Transcript.

Krebs, S. (2015). Einleitung: Zur Sinnlichkeit der Technik(geschichte). Ist es Zeit für einen »sensorial turn«? *Technikgeschichte, 82*(1), 3–10.

Krebs, S., & Van Drie, M. (2014). The Art of Stethoscope Use: Diagnostic Listening Practices of Medical Physicians and "Auto Doctors", *ICON: Journal of the International Committee for the History of Technology 20*(2), 92–114.

Kursell, J. (Ed.). (2008). *Sounds of Science-Schall im Labor (1800–1930).* Berlin: Max Planck Institut für Wissenschafsgeschichte.

Labelle, B. (2011). *Acoustic Territories: Sound Culture and Everyday Life*. New York, NY: Continuum.

Lachmund, J. (1994). *Der abgehorchte Körper: Zur historischen Soziologie der medizinischen Untersuchung*. Opladen: Westdeutscher Verlag.

Lachmund, J. (1999). Making Sense of Sound: Auscultation and Lung Sound Codification in Nineteenth-Century French and German Medicine. *Science, Technology and Human Values, 24*(4), 419–450.

Latour, B. (1986). Visualisation and Cognition: Thinking with Eyes and Hands. *Knowledge and Society: Studies in the Sociology of Culture Past and Present, 6*, 1–40.

Latour, B. (1987). *Science in Action: How to Follow Scientists and Engineers Through Society*. Milton Keynes: Open University Press.

Lethen, H. (2000). "Knall an sich": Das Ohr als Einbruchstelle des Traumas. In I. Mülder-Bach (Ed.), *Modernität und Trauma: Beiträge zum Zeitenbruch des Ersten Welkrieges* (pp. 192–210). Wien: Universitätsverlag.

Lynch, M. (1990). The Externalized Retina: Selection and Mathematization in the Visual Documentation of Objects in the Life Sciences. In M. Lynch & S. Woolgar (Eds.), *Representation in Scientific Practice* (pp. 153–186). Cambridge: MIT Press.

Mahrenholz, J.-K. (2008). Etnografische geluidsopnames in Duitse krijgsgevangenenkampen tijdens de Eerste Wereldoorlog. In D. Dendooven & P. Chielen (Red.), *Vijf continenten in Vlaanderen* (pp. 161–165). Tielt: Lannoo.

Mayer, K. (2011). Scientific Images? How Touching! *Science, Technology & Innovation Studies, 7*(1), 29–45.

Mody, C. C. M. (2005). The Sounds of Science: Listening to Laboratory Practice. *Science, Technology and Human Values, 30*(2), 175–198.

Mody, C. C. M. (2014). Essential Tensions and Representational Strategies. In C. Coopmans, J. Vertesi, M. Lynch, & S. Woolgar (Eds.), *Representation in Scientific Practice Revisited* (pp. 223–248). Cambridge: MIT Press.

Mundy, R. (2009). Birdsong and the Image of Evolution. *Society and Animals, 17*(3), 206–223.

Mundy, R. (2010). *Nature's Music: Birds, Beasts, and Evolutionary Listening in the Twentieth Century* (Ph.D. thesis, New York University).

Myers, N. (2008). Molecular Embodiments and the Body-Work of Modeling in Protein Chrystallography. *Social Studies of Science, 38*(2), 163–199.

Orr, J. E. (1996). *Talking About Machines: An Ethnography of a Modern Job*. Ithaca and London: Cornell University Press.

Pacey, A. (1999). *Meaning in Technology*. Cambridge: MIT Press.

Pasveer, B. (1989). Knowledge of Shadows: The Introduction of X-Ray Images in Medicine. *Sociology of Health & Illness, 11*(4), 360–381.

Pasveer, B. (2006). Representing or Mediating: A History and Philosophy of X-Ray Images in Medicine. In L. Pauwels (Ed.), *Visual Cultures of Science:*

Rethinking Representational Practices in Knowledge Building and Science Communication (pp. 41–62). Lebanon, NH: University Press of New England.

Pickstone, J. (2000). *Ways of Knowing: A New History of Science, Technology and Medicine.* Manchester: Manchester University Press.

Pickstone, J. (2007). Working Knowledges Before and After Circa 1800: Practices and Disciplines in the History of Science, Technology, and Medicine. *ISIS, 98*(3), 489–516.

Pinch, T., & Bijsterveld, K. (2012). New Keys to the World of Sound. In *The Oxford Handbook of Sound Studies* (pp. 3–35). Oxford: Oxford University Press.

Polanyi, M. (1983/1966). *The Tacit Dimension.* Gloucester, MA: Peter Smith.

Porcello, T. (2004). Speaking of Sound: Language and the Professionalization of Sound Recording Engineers. *Social Studies of Science, 34*(5), 733–758.

Rawlinson, A. (1923). *The Defence of London 1915–1918* (2nd ed.). London: Andrew Melrose.

Rice, T. (2013). *Hearing and the Hospital: Sound, Listening, Knowledge and Experience.* Canon Pyon: Sean Kingston Publishing.

Rice, T., & Coltart, J. (2006). Getting a Sense of Listening: An Anthropological Perspective on Auscultation. *The British Journal of Cardiology, 13*(1), 56–57.

Roberts, L. (1995). The Death of the Sensuous Chemist: The "New" Chemistry and the Transformation of Sensuous Technology. *Studies in the History and Philosophy of Science, 26*(4), 503–529.

Roosth, S. (2009). Screaming Yeast: Sonocytology, Cytoplasmic Milieus, and Cellular Subjectivities. *Critical Inquiry, 35*(2), 332–350.

Scarth, R. (1999). *Echoes from the Sky: A Story of Acoustic Defence.* Hythe: Hythe Civic Society.

Schafer, R. (1967). *Ear Cleaning: Notes for an Experimental Music Course.* Toronto: Berandol Music Limited.

Schafer, R. M. (1994/1977). *The Soundscape: Our Sonic Environment and the Tuning of the World.* Rochester, VT: Destiny Books.

Schirrmacher, A. (2016). Sounds and Repercussions of War: Mobilization, Invention and Conversion of First World War Science in Britain. *France and Germany. History and Technology, 32*(3), 269–292.

Schmandt, B., Jacobsen, S. D., Becker, T. W., Liu, Z., & Dueker, K. (2014). Dehydration Melting at the Top of the Lower Mantle. *Science, 344*(6189), 1265–1268.

Schrimshaw, W. (2015). Exit Immersion. *Sound Studies: An Interdisciplinary Journal, 1*(1), 155–170.

Schwartz, H. (2011). *Making Noise: From Babel to the Big Bang & Beyond.* New York: Zone Books.

Sennett, R. (2008). *The Craftsman.* New Haven, CT: Yale University Press.

Shapin, S., & Schaffer, S. (1985). *Leviathan and the Air-Pump: Hobbes, Boyle and the Experimental Life*. Princeton, NJ: Princeton University Press.

Snook, C. (1925). Automobile-Noise Measurement. *The Journal of the Society of Automobile Engineers, 17*(1), 115–124.

Stafford, B. M. (1994). *Artful Science: Enlightenment Entertainment and the Eclipse of Visual Education*. Cambridge: MIT Press.

Stangl, B. (2000). *Ethnologie im Ohr: die Wirkungsgeschichte des Phonographen*. Vienna: WUV Universitätsverlag.

Sterne, J. (2003). *The Audible Past: Cultural Origins of Sound Reproduction*. Durham: Duke University Press.

Suchman, L. (1987). *Plans and Situated Actions: The Problem of Human-Machine Communication*. New York: Cambridge University Press.

Supper, A. (2016). Lobbying for the Ear, Listening with the Whole Body: The (Anti-)Visual Culture of Sonification. *Sound Studies: An Interdisciplinary Journal, 2*(1), 69–80.

Te Hennepe, M. (2007). *Depicting Skin: Visual Culture in Nineteenth-Century Medicine* (Ph.D. thesis, Maastricht University).

Tufte, E. R. (1997). *Visual Explanations: Images and Quantities, Evidence and Narrative*. Cheshire, CT: Graphics Press.

Tyler, S. A. (1984). The Vision Quest in the West, or What the Mind's Eye Sees. *Journal of Anthropological Research, 40*(1), 23–40.

Van der Voort, A. W. M., & Aarts, R. M. (2009). Development of Dutch Sound Locators to Detect Airplanes (1927–1940). In *Proceedings NAG/DAGA, Rotterdam, The Netherlands, March 23–26* (pp. 250–253).

Vertesi, J. (2014). Drawing as: Distinctions and Disambiguation in Digital Images of Mars. In C. Coopmans, J. Vertesi, M. Lynch, & S. Woolgar (Eds.), *Representation in Scientific Practice Revisited* (pp. 15–35). Cambridge: MIT Press.

Volmar, A. (2012). *Klang als Medium wissenschaftlicher Erkenntnis: Eine Geschichte der auditiven Kultur der Naturwissenschaften seit 1800* (Ph.D. thesis, Universität Siegen).

Volmar, A. (2014). In Storms of Steel: The Soundscape of World War I and Its Impact on Auditory Media Culture During the Weimar Period. In D. Morat (Ed.), *Sounds of Modern History: Auditory Cultures in 19th- and 20th-Century Europe* (pp. 227–255). New York, NY: Berghahn.

Volmar, A. (2015). Ein "Trommelfeuer von akustischen Signalen": Zur auditiven Produktion von Wissen in der Geschichte der Strahlenmessung. *Technikgeschichte, 82*(1), 27–46.

Woolf, D. R. (2004). Hearing Renaissance England. In M. M. Smith (Ed.), *Hearing History: A Reader* (pp. 112–135). Athens: University of Georgia Press.

CHAPTER 2

Sonic Signs: Turning to, Talking About, and Transcribing Sound

Abstract This chapter focuses on processes of representing and sharing sound in the sciences. How have scientists, engineers, and physicians talked about sound and transcribed sound into legible signs? What did they do to ensure the acceptability and standardization of their verbalizations and notations? Why did embodied forms of notation survive despite a wider trend toward mechanical objectivity? And in what contexts did scientists become interested in the epistemological relevance of sound in the first place? This chapter also introduces most of the case studies and listening technologies in more detail.

Keywords Sonic signs · Representing and sharing sound · Embodied notation · Automatic registration · Mechanical objectivity

INTRODUCTION

In 1954, Lawrence N. Solomon earned his doctoral degree in psychology at the University of Illinois with a study on complex auditory stimuli. The stimuli he examined were underwater sounds. These were highly relevant in undersea warfare, giving sonar operators information on whether, for instance, battleships from their own country were coming home or enemy submarines were approaching. Solomon wanted to know how sonar operators made sense of such sounds, and he was especially intrigued by the informal "sonar vocabulary" the operators had developed to distinguish between and communicate about the sounds that they heard. The sonar men described sonic signs with words such as "heavy," "light," "bright," "dull," "hard," and "soft," making Solomon "suspect that <u>synesthetic</u> or <u>metaphorical</u> thinking is operative in this judgmental process." He assumed that the sonar operators had "intuitively turned for aid" to "these qualitative 'meaning' dimensions" to discriminate between the sounds (1954: 3–4).

Ultimately, Solomon's aim was to operationalize such meaning dimensions for psychological research. His dissertation contributed to the rise of the "semantic differential," a method for measuring meaning that was developed by Solomon's doctoral advisor Charles E. Osgood and his colleagues. The method asks test subjects to rate their attitude to a particular issue on a scale between two opposite adjectives—"heavy" and "light" in Solomon's example. It has been adopted in fields as diverse as psychology, acoustics, music, linguistics, business research, and political science to measure attitudes, opinions, values, and aesthetic perception.[1] The question of how to recognize and talk about sound thus informed the development of a test that is still widely used today.

The sonar men's meaning-making strategies also show the relevance of describing sound in the process of building knowledge on sound. Their synesthetic and metaphorical verbalizations of sound are instances of representing and sharing sound—the topic of this chapter. Other examples include the use of onomatopoeic terms; drawing lines, curves, and dashes; working with staves, keys, or notes; and having machines like spectrographs transform sounds into legible signs. Those who fostered listening for knowledge defined or redefined their interests as objects of sonic investigation, as issues that could be understood through practices of listening. To make their case, they not only tried to find ways of talking and writing about sound, but also of recalling sounds days, weeks, months, or even years after these had been audible. They tried to capture sound in terms that did justice to their questions and enabled them to share the sounds with their peers. They promoted strategies for analyzing sonic signs after the event. And they attempted to convince colleagues of, and initiate novices into, the sonic patterns they thought they could hear. These were the issues professionals and scholars working with sound dealt with and translated into practices that expressed both their sonic skills and ways of knowing.

This chapter covers both non-automatic, embodied practices of representing and sharing sound, such as verbalization and manual forms of notation, and the use of automatic, mechanical registrations of sound, such as audio recordings and sound spectrograms. This dual approach is important because in some fields, the development of manual notation has been closely associated with the availability of automatic recording technologies. The transcription of non-Western music, for instance, started to flourish only after the introduction of the phonograph (Stockman 1979), and naturalists' embrace of sound recording technologies as a way to bring bird sounds home gave new salience to questions of how to properly transcribe these sounds.

What notation systems did scientists, physicians, and engineers who listened for knowledge employ in their research and teaching, and where did those systems come from? How did existing and new forms of notating and recording sound co-constitute both the objects of research and the ideal researcher him- or herself? And under what conditions did non-automatic forms of representation survive alongside automatic forms? The crux of my answer will be that manual notation and other forms of embodied representation survived notably in situations where the required auditory knowledge depended on immediate, on-the-spot judgments and communication by trained experts with jurisdiction, as well as in situations where the gestural nature of manual notation had its own epistemological and didactic value.

In addition to discussing the relevance of representing and sharing sound for the process of making knowledge out of sound, this chapter unpacks the Sonic Skills project's core case studies by detailing the sources behind them. It also contextualizes the situations in which particular experts began to feel the need to listen to their objects of interest in the first place. A more systematic discussion of the purposes of listening, and of ways of listening in the sciences, will be presented in Chapter 3.

Hazy Sounds: Verbalizations and Descriptions of the Audible

Solomon's research on sonar was done during his summer job in San Diego as a psychologist at the United States Navy Electronics Laboratory in the early 1950s (Solomon 1954: 65). Sonar equipment enabled navy submarine operators to pick up and amplify sound signals in water, such as those produced by ship propellers. They had to distinguish relevant from irrelevant sounds and label the sources correctly by honing in on subtle changes in the sound's loudness, pitch, timbre, and rhythmic pulse. To understand these processes, Solomon wrote, one had to take into account not only the physical properties of the sound stimulus—intensity, frequency, complexity, and duration—but also what the stimulus meant to the person responding to it. He also argued that the relationships between "certain stimuli and their connotative associations" were "lawful" (Solomon 1954: 9).

To find those laws, Solomon designed a list of fifty pairs of attributes for the assessment of passive sonar sounds, such as heavy-light, smooth-rough, powerful-weak, hot-cold, green-red, masculine-feminine, clear-hazy, and mild-intense (Solomon 1954: 27). Many of these words came from lists of recognition cues employed by the sonar operators themselves. Solomon then asked fifty sonar men to listen to a selection of Navy recordings of passive sonar sounds, and rate each one on a scale of 1 to 7 for all fifty attributes. The greatest consistency in ratings was found in the pairs heavy-light, mild-intense, and clear-hazy (Solomon 1954: 43–45). Comparing these results with those published by Charles Osgood on other groups, Solomon concluded that notions expressing "clarity," such as clear-hazy, and "security," such as mild-intense, were most specific to what he called "sonar culture" (Osgood et al. 1957: 68). The clarity attributes expressed the ways in which sonar men selected "a coherent noise signal from a background of noise," and the security attributes how they discriminated between the most "heavily armed" vessels and less dangerous ones (Solomon 1954: 44, 48).

Of course, correctly interpreting the enemy's sound was often crucial for survival in war situations. From understanding the impact of acoustic shadows—hills, bushes, snow, and wind refracting or absorbing sound so that one might not hear an enemy attacking—to detecting and locating hostile artillery or uncovering secret atomic testing; sound could be the difference between life and death (Ross 2004: 275; Schwartz 2012: 573; Volmar 2012, 2013, 2014). The wide range of technologies to enhance listening in World War I was mentioned in the previous chapter, but occasionally, sound was even used to fake military actions. At the end of World War II, a US "ghost army" acoustically misled the Germans by suggesting an attack on one position while assaulting another. This sonic deception was achieved by replaying sound recordings of military operations—tanks driving, bridges being constructed, soldiers screaming—through loudspeakers mounted on "sonic cars" (Gerard 2002: 100–121, 277–292; Goodwin 2010:

41–43). Understanding sound and communicating about it comprehensibly, then, were extremely important for the military, and especially for sonar experts. Consequently, sonar operators received extensive training in sound detection and recognition, largely behind closed doors. Solomon, for instance, kept the exact sources of the sounds listened to by his sonar respondents secret for security reasons (Solomon 1954: 19).

In the worlds of the doctors, mechanics, engineers, and scientists we studied, unlike in the military settings, the sounds were usually not kept secret. However, just as much importance was attached to the creation of taxonomies of words and notation systems for sounds, intended to classify and standardize descriptions in order to consistently distinguish sounds from each other. That undertaking was far from easy. Drawing up a classification was one thing; having it adopted beyond a local culture like that of sonar quite another. The same applied to the codifications that mapped particular sounds onto particular problems. And the debates about classification, standardization, and systematic sound mapping tended to intensify even more once groups of experts shifted from unmediated listening to mediated listening, each with their own instruments.

As Jens Lachmund phrased it wonderfully when discussing the early nineteenth-century introduction of the stethoscope in medicine, codification systems "transformed fleeting auditory experiences into a world of communicable signs and meanings" (Lachmund 1999: 420). In medicine before the nineteenth century, listening to the body had only been possible by placing one's ear upon the body—something often felt to be inappropriate, especially if the patients were ladies. In the second half of the eighteenth century, an alternative arose: "percussion," a diagnostic listening to the sounds produced by knocking on the body. The stethoscope conveniently allowed doctors to physically distance themselves even further from the patient, while also enabling them to listen to bodily sounds with more focus through "auscultation." The resulting perceptual distance and isolation of the sounds of interest afforded, Tom Rice has argued, a process of "acoustic objectification" (Rice 2013: 126). The transition from the monaural stethoscope to the binaural stethoscope in the early 1850s helped to create a "enclosed aural pathway" between the doctors' ears and the patients' bodies, the sounds of which could be amplified with electronic stethoscopes from the 1920s on (Van Drie 2013: 176).[2]

Listening for knowledge with the stethoscope was not only dependent on acoustic objectification and an enclosed aural pathway, however. Just as important was the communication about what doctors were aurally attending to. Working in Paris in the early nineteenth century, René Laennec presupposed a direct relationship between particular pathological conditions and certain "pathognomic signs" he heard with his stethoscope, which he classified as signs of the voice, respiration, rattles (fluid-induced alterations in respiration), and circulation (Lachmund 1999: 425). To distinguish between these sonic signs, he made a wide range of comparisons with everyday sounds such as "the satisfied purring of cats being petted," "the tinkle of weapons during military exercises," or "the sound produced when a string of bass is beaten with a finger" (Laennec cited in Martin and Fangerau 2011: 303, and in Lachmund 1999: 425–426). Laennec did not only invoke everyday soundscapes, however; he also used visual imagery, such as "bubbles like those which are produced by blowing with a pipe into soapy water," to capture what was happening inside diseased bodies (Laennec cited in Lachmund 1999: 426) He considered the hospital setting indispensable for acquiring stethoscopic skills, not only because of the number and diversity of patients available for practicing but also because of the opportunity to check assumptions retrospectively through autopsy—again seeking visual analogies. Colleagues working in the same tradition, such as Jules Fournet, created increasingly fine-grained and multilayered classifications drawing on the tradition of botanical taxonomies, and added "synoptic tabular visualization of auscultation sounds and their meanings" (Lachmund 1999: 428).

In contrast, the Viennese doctor Joseph Skoda postulated that physics and acoustic phenomena, such as reverberation and amplification, were also needed in order to establish a convincing relationship between sounds and pathologies. This attention for the acoustic milieu of sounds resulted in a less detailed classification, as not all deviating sounds could be linked directly to particular health problems. Skoda still employed metaphors, and also used onomatopoeic words such as "*tik-tak, tom-tum*" and "*dohm-lopp*" for heart sounds (Martin and Fangerau 2011: 304), but he did not seek to create a complex taxonomy assigning one sound to one cause. Instead, he invited doctors to mimic particular sounds with their own body in order to understand the acoustic phenomena that helped to define the sounds of pathological conditions. In addition, he developed a system of "differential diagnosis," in which signs were not universal and positive references to particular pathological conditions, but starting points for a negative strategy based on excluding certain interpretations. An audible intensification of a patient's voice, for example, might be the acoustic effect of "thickening of the pulmonary tissue," but its specific cause—the disease behind it—had to be found by excluding alternative interpretations through anamnesis and other techniques (Lachmund 1999: 432). This strategy, Lachmund argues, was embedded in an experimental laboratory tradition of medical research in the German-speaking area, where doctors had access to fewer patients than did French doctors with their large hospitals. In France, tacit sonic skills could be transmitted within an apprenticeship context; learning sounds from textual descriptions alone was much harder. The alternative was to reject Laennec's and his followers' straightforward tables of sonic signs and related conditions in favor of differential diagnosis.

In the context of twentieth-century German automotive engineering, a predominantly tacitly transmitted system of sound mapping developed, as Stefan Krebs has shown by analyzing German automotive trade journals, car mechanics' handbooks, the Volkswagen and Robert Bosch company archives, and interviews with former car mechanics. Just as in French medicine, an apprenticeship tradition enabled a culture of locally taught, and locally understood, verbal descriptions of sound to arise in German car mechanics. In the early 1930s, German car mechanics succeeded in having their trade legally protected in craft guilds (Krebs 2012a: 200). The move responded to public concerns about the quality of car repair, and helped them to demarcate their work from ad hoc mechanics—but it also secured them jurisdiction over the diagnosis of alarming car sounds at the expense of motorists. Motorists' handbooks had long instructed drivers to monitor and diagnose their cars' functioning by listening while driving (Bijsterveld 2012: 161; Bijsterveld et al. 2014; Krebs 2012a: 99–102; 2013: 92); now, car mechanics increasingly labeled motorists as either "noise fanatics" or "noise phlegmatics." Noise fanatics brought their cars to the garage in response to the slightest, often wrongly understood sounds, whereas noise phlegmatics kept driving even if their cars were simply screaming for help (Krebs 2012b: 96). The perfect motorist, mechanics believed, would have just enough auditory knowledge to identify sounds that signified car problems, bring his car to the garage, and leave the actual diagnosis to the mechanic.

With a three-year apprenticeship ideally culminating in a journeyman's certificate and a four-year on-the-job curriculum for the master craftsman's diploma, mechanics were taught in a learning-by-doing context. On the shop floor, novices were supposed to develop not only technological knowledge and mechanical proficiency, but also sensory skills. Two short series of articles published in brochures and a journal for apprentices in 1956 and 1965 explained the importance of these sensory skills in diagnosis. The "expert eye" expressed itself in an "intuitive gaze" that enabled the mechanic to "notice every deviation from the 'normal picture' ... such as oil slicks or rust stains" right away. The "expert nose" would be able to smell gasoline or burned cables, and tasting liquids could be useful, although potentially dangerous to the mechanic's health. The "expert ear" was crucial as well—"probably even more important than the expert eye," one author had it. "Listening to recognize the actual problem, listening in, with a listening rod or just a screwdriver, to locate it—that is real diagnostic practice" (Anonymous 1956, 261–263, cited in Krebs 2014a: 360–361; 2014b: 80). This was anything but simple to learn, because the "real art of 'listening' to automobiles only starts where one complex sound dissolves into many single sounds and the mechanic's ear will be able to connect a particular source of noise with each one of them" (Anonymous (Teil II), 1965: 235, cited by Krebs 2014a: 361; 2014b: 81).

An experienced mechanic, in this view, would be able to distinguish a problematic, discontinuous "prr-prr-prr" from a reassuring, continuous "prrrrrr" and know that it signified an issue with one of the cylinders. He would be able to key such sounds into his theoretical knowledge of how subsystems in the car affected each other, and might check, for instance, whether a particular sound disappeared or changed in pitch once specific parts were separated from the engine and drive train or the gears were moved up and down. Analyzing the sound's rhythm through its periodicity was another significant strategy. If the sound was audible only every second revolution rather than every revolution, the fuel pump might be broken. For the act of listening, mechanics used their hands to shield their ears from disturbing noises, or employed tools such as rods, screwdrivers, and stethoscopes. Some of these stethoscopes had exchangeable probes—tips or bells—to vary between listening to large and small spots, and were exclusively used to reach specific parts of the car such as the engine block or individual bearings. One additional instrument, a device for testing the circuits in generators and starting mechanisms, had an auditory display in the form of a buzzer to which mechanics could listen with headphones (Krebs 2014a: 362).

Sensory skills and diagnostic listening were not so highly valued in all circles of car mechanics, however. The United States, for instance, witnessed a similar "repair crisis" to Germany's in the 1930s, but a very different response prevailed. Whereas German mechanics had successfully joined forces to rhetorically undermine motorists' auditory diagnostic capabilities and claim diagnosis through listening as their exclusive trade expertise, US motorists effectively turned against car mechanics. They questioned the workmen's choices—were their expensive repairs really necessary, or based on subjective assessments favoring their wallets? Car mechanics responded by trying to enhance the transparency of their actions, resulting in forms of diagnosis that used meters such as "the voltmeter, ammeter, and ohmmeter, as well as various compression or vacuum gauges" instead of the senses (Krebs 2012b, 2014a: 365). I will return in Chapter 4 to this move towards visual diagnostics in US automotive repair.

In the German automotive world, diagnostic listening survived much longer, though mechanics and automotive engineers invested less sustained effort in refining and ordering the descriptors of car sound than the medical world had done for bodily sounds. To be sure, the German automotive press published lists, tables, and fault trees describing sounds and their causes. Onomatopoeia were little used, but terms such as whining, rumbling, hammering, pinging, and knocking were ubiquitous, at times further specified as "metallic knocking," "damped knocking," or a "muffled clang" (Anonymous 1932: 81–82, cited by Krebs 2012b: 88). One article from the late 1950s explicitly listed twenty-five such suspicious sounds. Several authors, within and outside Germany, acknowledged that the automotive world used different sound labels for the same phenomenon (Krebs 2014a: 363; Zwikker 1934: 70). In turn, one particular sound, such as the knocking of engines, at times compared to the muffled rumble of a kettledrum, could be a sign both of machine wear and of gasoline problems (Bijsterveld 2007: 15). Most authors did not worry too much about such ambiguities because they anyway regarded written accounts as unhelpful for learning diagnostic listening—experience was the key to the mechanic's training. The automotive world could afford to neglect the construction of unambiguous descriptions either because diagnostic listening was firmly located within the jurisdiction of mechanics and their tacitly transmitted strategies for connecting textbook knowledge to sensory observations on the shop floor, or because it had simply been replaced by meter reading under pressure from critical customers.

The medical world, in contrast, continued to discuss the disambiguation and standardization of descriptions of sound, increasingly bearing in mind the need for international scholarly exchange. In the second half of the nineteenth century, a German physician described the French medical classifications of sound as "Chinese" in their complexity and incomprehensibility (Niemeyer 1868: 19, cited in Martin and Fangerau 2011: 304). The suggested alternatives included differential diagnosis as well as systems for graphic notation of heart sounds, which I will discuss in the next section. In respiratory medicine, the "universal semantic" that would standardize the descriptors of bodily sounds and their meaning became an elusive holy grail (Reichert et al. 2008: 2). The International Lung Sounds Association, established in 1975, fostered the "exchange of ideas and experience" in respiratory medicine, and hoped that "comparisons of methods of recording, analyzing, and describing lung sounds" would "reduce ambiguity." Accordingly, it instituted a Committee on Lung Sound Nomenclature, but did not expect that "a new and improved set of terms will be agreed upon and recommended for instant acceptance by the medical profession."[3] Sure enough, in 2000, a group of physicians and scientists collaborating on software for Computerized Respiratory Sound Analysis (CORSA) still found it necessary to first collect and define over 160 relevant terms. Among these was the "crackle," an "[a]dventitious, discontinuous, explosive sound occurring usually during inspiration" (Sovijärvi et al. 2000: 600).

Despite such attempts at international standardization, doctors still referred to everyday sounds of their own choice when introducing students to the mysterious world audible with the stethoscope. We know this from the studies undertaken by Anna Harris and Melissa Van Drie. Harris, a medical graduate turned anthropologist, carried out an ethnography of listening practices at a Melbourne hospital medical school for five months in 2013, and observed training in physical examination at the Maastricht University Skills Lab the year after. In Melbourne and Maastricht, she performed a total of seventeen formal semi-structured interviews with doctor-teachers and students, as well as talking informally to them, to nurses, and to other hospital staff. She recorded the interviews and her auditory observations of hospital sounds, and made drawings of what she saw teachers and students themselves drawing. Van Drie selected the most widely reprinted and widely cited British, American, and French textbooks on physical examination and lung auscultation guides from 1950 to 2010, and searched medical libraries and archives for evidence on other media employed to teach stethoscope listening. As noted in Chapter 1, both Harris and Van Drie focused on the teaching of the respiratory cycle, but without neglecting the training of students to listen to the cardiovascular system.

Hardware for student instruction, Van Drie (2013; Harris and Van Drie 2015) shows, evolved from collective stethoscopes with multiple earphones attached to one bell (1880s), via electronic collective stethoscopes (1920s) and plug-in electronic stethoscopes attached to a lecture hall broadcasting system (1960s), to today's mannequin simulators, and from gramophone recordings as textbook appendices (1930s), via audio cassettes to be listened to without (1950s) or with a stethoscope (1970s), to present-day compact disks and mp3 files, enabling students to listen to exemplary bodily sounds. Nevertheless, metaphors remained helpful. They were both culturally and historically specific: Dutch medical students nodded in assent on hearing an instructor explain that pleural friction rubs sounded like "feet crunching in snow," but that would hardly have been self-evident to students living closer to the equator. An Australian professor had long been accustomed to describing the sounds of pleuritic rubs as squeaking leather, but acknowledged that this ceased to be effective when Melbourne trams stopped using leather suspension components. The same professor, however, did not yet wholeheartedly endorse the more recent description of fine crackles as Velcro being pulled apart (Harris and Van Drie 2015: 103). And Sarah Maslen (2015) has noted that physicians still employ not only onomatopoeic signifiers, such as "lub," "dub," and "lub-di-dub" for the first, second, and third heart sound, but also metaphors: the fourth, pathological heart sound is "like a hose directly hitting against a bucket." Such descriptions, Maslen believes, help students to memorize instructions, and—bearing in mind George Lakoff's and Mark Johnson's work on the metaphors we live by—"to structure understanding" (Maslen 2015: 61, 64).

The contexts into which descriptions, classifications, and codifications of sound are intended to intervene are thus crucial for their form and for their acceptance as relevant expertise. I will return to this issue later in the chapter, but let me first discuss yet another way of describing sound. As they seek words for sound, doctors, engineers, and mechanics have occasionally used musical metaphors. Laennec referred to bass strings, 1920s engineers in the automotive world to kettledrums—at times picturing a noisy engine as an unruly big band (Bijsterveld and Krebs 2013: 19). The actual use of musical notation was rare, however, even though Laennec used it to put heart sounds on paper (Lachmund 1999: 428; Martin and Fangerau 2011: 305). In ornithology, by contrast, musical staff notation was a dominant practice at least at the dawn of the field as academic discipline, around 1900. One key to understanding this difference between the sciences is the conceptualization of the sounds being studied. Initially, bird naturalists saw themselves as studying not bird *sound*, but bird *song*, a choice of terms that expressed their early inclination to classify bird vocalizations as a form of music. It therefore seemed perfectly natural to notate bird sound in musical staff, as Joeri Bruyninckx (2013) showed in his study of American, British, and German ornithology.

Bruyninckx systematically analyzed key naturalist and ornithological journals that entered the scene between 1880 and 1980, *The Auk, The Condor, The Wilson Bulletin, Journal of Field Ornithology, Ibis, British Birds, Journal für Ornithologie, Animal Behaviour,* and *Behaviour.* He also traced the rise of the bird-sound recording archives, studying in most detail the earliest of these—the Macaulay Library at Cornell University in New York, the British Library of Wildlife Sounds in London, and the Tierstimmenarchiv at the Humboldt University in Berlin. Additional sources were the autobiographical accounts of pivotal players in the world of bird studies in the past, and oral interviews with their counterparts in the present.

ARTFUL DRAWINGS: MUSICAL AND GRAPHIC NOTATION

In the early decades of the twentieth century, most ornithologists used musical staff notation, often in combination with syllabic notation of onomatopoeic expressions. The British ornithologist Walter Garstang, for instance, wrote syllables such as "sip, sip, sip, see! Tee, tew, wee, tew! Wit-ty, wit-ty, wee-wee, wee tew!" below his musical score of the Willow-wren (Garstang 1922, cited in Marsden 1927: 342). Some even expanded such melodic musical scores into "four-part" harmonies that could be played on a piano or other harmonic musical instruments (Mundy 2010: 181). In fact, composer Cornell Schmitt and naturalist Hans Stadler advocated musical staff notation as the scientific method par excellence, on the grounds that its mature conventions provided a "precise and scientific way of comparison" (Schmitt and Stadler 1913: 394, cited in Bruyninckx 2013: 40). Musical training, claimed biologist Henry Oldys, allowed naturalists "to observe important features that are quite certain to escape the attention of one whose musical ear has never been cultivated" (Oldys 1916: 20, cited in Bruyninckx 2013: 39).

This reliance on musical notation was fueled in part by the expectation that studying birdsong might disclose the origins of human music, a notion that itself drew on the Darwinian argument that birdsong, like plumage, emerged from the mechanism of sexual selection. In the 1910s, opinion in ornithological journals began to diverge on whether birdsong was indeed a function of sexual selection, or rather of natural selection or simply imitation. It was also unclear whether birds could themselves be ascribed aesthetic sensitivity. In the 1930s, the rise of ethology pushed aside the issue of the bird as a proto-artist: biologists such as Konrad Lorenz and Nikolaas Tinbergen "conceptualized animal behavior instead as compulsive, functional and automatic" (Bruyninckx 2013: 45). Their approach marked the culmination of another gradual shift in ornithology since the 1880s, from a focus on collecting, describing, and classifying *dead* birds for fauna taxonomies to a focus on studying birds *live*, especially in their natural settings.

In line with this new type of interest in birds, it was in a 1904 field guide for wild bird life that the American amateur naturalist Ferdinand Mathews noted his difficulties with musical notation when trying to describe a Bobolink song. The result was a remarkable transcript.[4] Mathews admitted he had "never been able to 'sort out' the tones as they passed at this break-neck speed," so that "the difficulty in either describing or putting upon paper such music is insurmountable" (Mathews 1904: 49, cited in Bruyninckx 2013: 29–30). In Bruyninckx's words, the drawings that Mathews proposed instead "started off in a traditional grid of quarter and eighth notes but quickly oscillated beyond the conventional dimensions of relative time and pitch," showing notes "wildly bouncing and receding back in time" (Bruyninckx 2013: 30). Another American amateur naturalist, Aretas Saunders, argued that musical notation "has been made primarily for the recording and rendering of human music and birds do not usually sing according to such standards." Soon after, he rhetorically asked whether ornithologists should "change such a song in order to make it fit our method?" Would such a procedure express "scientific accuracy" (Saunders 1915: 173, and 1916: 104, cited in Bruyninckx 2013: 46–47)? Notating bird sound in terms of Western music was increasingly seen as endangering a scientific understanding of bird vocalization.

The American naturalist Lucy Coffin promoted the use of "new musical scales akin to Chinese or Gregorian ones" as well as a wider range of instruments, "including xylophone, banjo, zither, bassoons, and piccolo," to represent bird sound more precisely (Coffin 1928, cited in Mundy 2009: 208). British poet William John Murray Marsden wrote that he was looking for "musico-ornithologists" capable of effectively notating the song of birds, as the "diatonic notes, on a keyed instrument, are 'not a bit like' what one hears from the coral beak." He had been able to "whistle and fiddle" a few cries and phrases himself, but he suspected that "with a few pretty well recognised exceptions, no bird's song—or speech, or whatever it is—can be expressed in tones of our musical notation. And I want somebody, now that, as I understand, our diatonic system is undergoing Promethean and Protean experiences in younger hands, to be inspired to bring together again the birds and ourselves" (Marsden 1927: 339–341). Marsden was probably referring to the composers of his time and their experimentation with new notation forms. Composers such as Ferruccio Busoni, Luigi Russolo, and Henry Cowell aimed to free contemporary music from the limits of traditional musical notation and enable the notation of microtones, as well as new rhythmic and harmonic complexities that better expressed modern urban life and composers' imaginations (Busoni 1962/1907; Russolo 1986/1916; Cowell 1932).

The type of graphic notation that Aretas Saunders suggested for bird vocalizations similarly permitted the representation of microtones.[5] Saunders did away with conventional staves and their relative organization of pitch and time, instead plotting pitch in terms of absolute frequency—even though this frequency was still indicated using musical notes. He recorded duration, too, in absolute terms, that is, in seconds. This met with fierce opposition. For the musician and naturalist Robert Moore, a dedicated promoter of musical notation in the American Ornithologists' Union, meter and rhythm were such crucial elements of birdsong that their registration must not be left out (Bruyninckx 2013: 47). Clearly, opinions on the best methods of representation went hand in hand with conceptualizations of the object under study.

But in the ears of some ornithologists, including Saunders, documentary infidelity was not the only problem: they also deplored the exclusion of people unable to read music.[6] Their alternatives did not go as far as

Mathews's notation, but were truly graphical or combined graphics—waves, straight lines, dots, dashes—with more traditional syllables to capture and evoke the sounds of birds. One advocate of such systems, the Canadian zoologist Bill Rowan, even promoted a shorthand script that aimed less for accuracy than for a "simplicity, plasticity and adaptability" allowing non-musicians to read and notate bird sound (Rowan 1925, cited in Bruyninckx 2013: 50). This was particularly important for a field science that depended on the input of amateurs as much as on that of professional ornithologists.

The issue of the best notation system for bird sound had much in common with to the problem of transcribing folk music and non-Western music. In musicology and ethnomusicology, the debate took off soon after the introduction of the phonograph for recording folk and non-Western music in the early 1890s in the United States and Europe. Among those involved were folk-oriented composers Béla Bartók and Zoltán Kodály, as well as music psychologists and ethnomusicologists (Stockmann 1979: 207–208). Highly influential was the 1909 "Proposal for the Transcription of Exotic Melodies" by Otto Abraham and Erich von Hornbostel. These two German scholars welcomed the rapidly expanding collections of phonograph recordings of non-Western music, to which von Hornbostel contributed as curator of the Phonogramm-Archiv in Berlin (Mundy 2009: 218). They worried, however, that these collections would remain "dead capital" if ethnologists and musicologists did not create a standardized form of notation enabling them to compare the recordings (Abraham and von Hornbostel 1909: 1). They did not want to end up like linguists, who had developed so many different diacritical signs for phonological work that—in Abraham and von Hornbostel's view—they had seriously complicated their lives as researchers.

The transcription of non-Western languages, so the Americanist Brian Hochman has shown, had been a hot issue ever since ethnologist Franz Boas claimed, in his 1889 essay "On Alternating Sounds," that the Western ethnographic ear perceived alternations in the pronunciation of certain indigenous languages differently from the way native speakers themselves did. The admission of this "sound-blindness" and the wish to preserve languages resulted in enthusiasm for new orthographical techniques in literature and an archival shift to phonographic recording in the study of language and culture (Hochman 2010: 531–533). The phonograph could record languages "exactly" as the natives uttered them, and reproduce music as it had once been made. As the American music psychologist and ethnomusicologist Benjamin Ives Gilman wrote: "It can be interrupted at any point, repeated indefinitely, and even within certain limits magnified, as it were, for more accurate appreciation of changes in pitch, by increasing the duration of the notes" (Gilman 1891, cited in Hochman 2010: 541). The phonograph offered Gilman what Hochman calls a "culturally sealed container" for collecting and comparing "auditory specimens" free from field distractions (542).

Abraham and von Hornbostel fully acknowledged that Western notation did not do justice to non-Western music. At the same time, they wanted to modify the existing system as little as necessary, given that innovations in printing were costly and novel forms of notation were hard to learn and remember. For these reasons, they rejected systems that required extra lines beyond the usual five, something proposed by Gilman in 1908 and the composer Busoni in 1907. Graphic notation might be useful to show the curves of a melody, they conceded, but in principle notation should be done in musical staff in a simple key close to the original pitch, if necessary with a system of signs that combined sharp and flat to indicate recurrent deviations and + and − for tones between the half tones of the Western system. If pitch was unclear or noise-like, staff without heads was useful. Abraham and von Hornbostel proffered an extensive table of signs for exotic ways of phrasing, but continued to struggle with the notation of timbre, a classic problem in conventional music notation. The table used instrumentation to express tone color and had very few signs for ways of playing that evoked particular timbres, such as ° for flageolets. In addition, they endorsed Alexander J. Ellis's system of cents, introduced in 1885, which enabled absolute differences in pitch, or frequency expressed in "Herz," to be translated into relative distances between tones. One hundred cents stood for a half-tone difference and 700 cents for the seven half tones of a fifth; any number in between expressed microtonal deviations (Abraham and von Hornbostel 1909: 19). This plethora of suggestions indicates that, whereas the documentary fidelity of the phonograph was praised in ethnomusicological settings, graphic notation was far more contested as a method of standardization.

Standardizing notation was also an issue in dance. A short excursion into the debates in this field will prove helpful for my discussion of embodied versus mechanical notation in knowledge making in the penultimate section of this chapter. Folk-dance researchers notated dance using verbal descriptions and Western music notation combined with symbols expressing the movements of hands and feet (Douglas 1937: 114–116; Kenworthy Schofield 1928: 23–24). In 1935, on the occasion of a folklore conference and festival in London, the British diplomat and folklore researcher Rodney Gallop urged international agreement on a "formula." Not that such a formula could replace description; it "would give only an approximate idea of each dance, but it would enable the points in common between any two or more dances to be quickly picked out and the Highest Common Factor, to borrow a mathematical term, to be established between them" (Gallop 1935: 79). Several colleagues agreed, although one, the Oxford archaeologist and professor of ancient history John Lynton Myres, added that in creating such a nomenclature, folklorists would encounter the same problems of classification as ornithologists and zoologists had done (Myres in Gallop 1935: 82). They would still have to choose between the many systems used so far: verbal description; musical notation; musical notation enriched with words, figures, and steps; and the graphic notation of foot movements (Caravaglios 1935, 129–130).

Another commentator on Gallop's lecture, the Austrian folklorist Richard Wolfram, rejected the very idea of a truly universal system, as each type of dance required its own notation. Moreover, one could only really learn a dance by observing and interviewing the performers themselves (Wolfram in Gallop 1935: 82). And organologist Curt Sachs, also present at Gallop's presentation, warned that a notation system focusing on the component parts should not stop folklorists from approaching dances "as a whole." Instead of recording dances through a system of symbols, he suggested, they should "try the experiment of photographing a dance in the dark, with lights attached to the heads and limbs of the dancers." This would record the curves described by the dancers' bodies, from which the necessary comparisons and conclusions could be drawn (Sachs in Gallop 1935: 83).

Sachs thus favored the mechanical registration of dance, in the form of an abstraction of movements by observing the lines of the lights, whereas others preferred the lines of manual graphic notation. From the late nineteenth century, systems of manual graphic notation were also suggested in medicine, as an alternative to the existing wide variety of verbal descriptors. One of these systems applied metrical signs used in poetry (Martin and Fangerau 2011: 305). As historian Lisa Gitelman has shown, converting "aural experiences into inscribed evidence" beyond traditional script was a broader nineteenth-century preoccupation. One of Gitelman's examples is the rise of shorthand systems, or "phonography," in this case referring to the representation of "the sounds of speech on paper" (1999: 185). In medical practice, however, proposals to use graphic notation for plotting medical listening experiences once again led to a profusion of different sign systems.

For the case of German-language heart medicine, the medical historians Michael Martin and Heiner Fangerau (2011, 306ff) have explained that physicians tried to remedy the subjectivity of these codes around the turn of the twentieth century by introducing semi-manual notation systems. For example, they attempted to discipline the physicians' jottings through the use of a pre-formatted notation scheme. Because that still relied on individual physicians' observations, others proposed devices for the mechanical registration of sound that would be entirely independent of doctors' individual perception. These registration machines transformed sound waves into visual curves via electric signals or through a sensitive membrane, building on earlier work by scientists in acoustics and psychophysiology and following the wave-oriented visualization conventions of science at that time. Despite debates about the accuracy of these systems (some notation systems affected the results), their basic assumptions were readily accepted in medical laboratories. Proponents argued that such registrations could show details beyond human observation and enabled comparison both across registrations and across observers.

Clinicians were less enthusiastic. Rendering their observations open to inspection by others, including patients, potentially undermined the exclusivity of their expertise, and thus threatened to weaken their professional jurisdiction. They also believed that the promised precision of the machines would be difficult and time-consuming to attain. The stethoscope, in contrast, not only assisted the doctor in a quick and efficient check of the patients' health, but also functioned as a token of the physician's expertise and experience, as a symbol of his clinical habitus and social standing. Added to that, the stethoscope lent itself to being embedded in the important moments of direct contact between doctor and patient. If the stethoscope provided some physical distance, it also helped to keep physical examination, diagnostic work, and talking to the patient within a single sensory and temporal frame.

SAVE THE PATIENT: SURVIVAL
OF EMBODIED REPRESENTATION

From the late nineteenth century onward, ornithologists gradually embraced mechanical registration of bird sound by the cylinder phonograph. Yet even though several ethnologists, ethnomusicologists and zoologists incorporated Thomas Edison's phonograph in their work quite soon after its introduction to the market in 1878 and Emile Berliner's gramophone in 1887, ornithologists were not among the early adopters. Only caged birds could be brought close enough to the recording horn, and thus create the amount of acoustic energy, for their vocalizations to be recordable. The obsession of ornithologists like Saunders with achieving accurate and intelligible forms of manual notation may also have inhibited the growth of interest in mechanical recording, as Joeri Bruyninckx has suggested (2013: 49–50). The phonograph became far more relevant for ornithologists in the second half of the 1920s, with the rise of microphones and their capacity for electroacoustic amplification. This enabled the recording of birds where it mattered most: in their natural environment (Bruyninckx 2014: 44). It also coincided with the heyday of the ideal of mechanical objectivity, the automatic registration of phenomena without interference by the embodied subjectivity of researchers.

Indeed, Joeri Bruyninckx (2013) has shown, recordings were initially applauded for their ability to help overcome the subjectivity and imprecision of manual field notation (in other words, for their mimetic function). Sound recordings were also praised for their assistance in recalling and collecting bird sound (mnemonic function), in teaching novices the auditory recognition of particular bird species (didactic function), and, through their aesthetic appeal, in tempting a wide audience of lay birdwatchers to contribute to ornithology (alluring function). By the end of the 1920s, the broadcasting industry was fueling the appropriation of sound recording equipment, including sound cameras, in field work by hiring naturalists to make recordings for radio use (Bruyninckx 2012: 133; 2015: 349–350). Yet mechanical recording merely postponed the problems of manual notation and its standardization to the research phase after returning from the field. It resulted in a *repeatability* of listening that, as we have seen, also articulated the shortcomings of classical musical notation for bird sound and made ornithologists ask for a notation that was both enriched and systematized (Bruyninckx 2013: 49).

In the 1950s, the use of spectrography to visualize the frequency and intensity of sound across time started to dominate ornithological research. Many ornithologists welcomed mechanical visualization in the form of sound spectrograms, as it enabled an interesting focus on short calls and phrases in bird sound, creating "stills" of fragments of even the most rapidly alternating bird sounds. As Rachel Mundy has stressed, spectrographs also helped ornithologists to cope with the fact that bird sounds "often occur outside the human hearing range" (Mundy 2010: 180). Moreover, sound spectrograms permitted levels of standardization and distribution—or, in a Latourian sense, inscription, synoptic presentation, superimposition, and translation on paper—that they thought were more difficult to establish for manual notation. This increasing preference for mechanical visualization occurred not only in ornithology, but also in many other sciences of sound, such as acoustics, psychoacoustics, and phonology (Rieger 2009). That trend will be discussed in more detail in Chapter 4.

Why, then, did non-automatic, embodied registration practices survive despite the general trend towards mechanical objectivity and automatic registration? Even the heroes of ethnomusicological recording Abraham and von Hornbostel had to admit in 1909 that making mechanical recordings was not possible in every single situation. When it was not, manual notation could still be useful, provided it was done critically and with great care. Such care implied first learning to sing or play the music oneself, and then having an exceptionally musical member of the culture under study—musical in the terms of that culture—check whether the researcher had grasped the music correctly. Only after these steps should the music be translated into notation. Prepared in this way, manual notation could be highly instructive: the process of transforming the act of music-making into manual notation would enable the researcher to understand issues such as complex drum techniques and intricate melodic structures from within (Abraham and von Hornbostel 1909: 15). A similar argument had been made about dance by folklore specialist Wolfram in the mid-1930s, though his interest was not the opposition between manual and mechanical but between idiosyncratic and standardized notation. In both cases, the idea was that forms of manual notation linked to music-making or dancing by the researcher him- or herself would enhance the quality of the notation—not only in its mimetic dimension, but also in terms of analytical depth. Moreover, manual notation was a relatively slow process. As ethnomusicologist Helen Heffron

Roberts claimed in 1931, this created opportunities for questions to arise, to the singers for instance, which would not occur when making phonographic recordings (Brady 1999: 73).

It is almost as if these ethnomusicologists and folklorists had read recent work by the British anthropologist Tim Ingold (2007). In his comparative anthropology and history of the line, Ingold distinguishes manual, gestural sketching from printing because it embodies a way of knowing that is, in his view, more true to how people experience their environment than mechanical printing can be: in real life, "we perceive the environment not from a stationary point ... but in the course of our movement along what [James] Gibson calls 'a path of observations' ... In the freehand sketch, the movement of the observer relative to a stationary feature is translated into the movement of the line depicting that feature relative to a viewer who is now stationary" (2007: 166). It was in the connection between bodily movement—learning to sing, play, or dance to particular music—and the subsequent manual notation of music that Abraham, von Hornbostel, and Wolfram situated a particular understanding, just as Ingold does. Even if they did not make it explicit, they seem to consider this understanding to be tacit yet transferable through its embodied transcription. For them, the epistemological value of manual notation depended on its *sensory calibration* in embodied music-making.

Another set of explanations for the survival of manual notation and other forms of embodied representation revolves around issues of urgency, efficiency, and dependency. Just as ethnomusicologists did not always have their recording equipment on standby, ornithologists sometimes had to rely on shorthand scribbles, for instance when an interesting bird made itself heard just as the recorder batteries ran out. In those cases, immediate manual notation was, and is, both urgent and highly efficient.

But manual notation has also been found efficient in situations of a different kind. In one of the systems that was introduced in the late nineteenth century and has survived until today, lung sounds are notated in terms of their occurrence within the respiratory cycle, usually as ascending and descending lines denoting inhalation and exhalation. "The length of the line indicates variations in the duration of the breath sound; the pitch (high or low) is represented by the line's angle; intensity (loud or soft) is illustrated by the thickness of the line" (Van Drie 2013: 182). In another system, a variety of dots, squiggles, and waves are sketched onto a two-dimensional outline of the lungs and shared in learning situations (Lachmund 1999: 428). As Anna Harris has shown in her ethnographic study of a medical school in Melbourne, doctors training novices in stethoscopic listening often create these lines in the air by gesticulating with their hands while mimicking sounds, or make "drawings of respiratory rhythms and heart murmurs" with whiteboard markers to evoke particular sounds and direct students' attention to them (Harris and Van Drie 2015: 103). Doctors also encourage students to carry out "auto-auscultation," touching and listening to their own bodies in order to grasp their teachers' references to sound (Rice 2013: 19, 137ff). By tapping on their thighs, for instance, they may understand what it means to hear and feel a "dull sound," or, to learn "to be affected" (Harris 2015: n.p.; 2016: 46).

Other embodied forms of representing sound survived as well. In the guild-like context of German car repair shops, as we have seen, exchanges of verbal descriptions of sound remained important for the analysis of car engine problems, both between experienced mechanics, and between master and apprentice (Krebs 2012b, 2014a). And even vocalizations of sound did not entirely disappear. An example comes from Alexandra Supper's research on the sonification community (2012). She attended three conferences organized by the International Community for Auditory Display (ICAD); interviewed its founding father Gregory Kramer and thirty-three other sonification practitioners between 2008 and 2011; studied dissertations, conference proceedings, and journal articles on sonification; attended sonification performances, talks, and workshops; and experimented with sonification herself using the audio synthesis programming language SuperCollider. Supper mentions Thomas Hermann's vocal sonifications of EEGs, in which data are deliberately made to sound like human vowel sounds in order to enable "data karaoke," a way of mimicking sonified data with one's own voice. This is particularly useful, Hermann claims, when sonification experts and clinicians collaborate on the interpretation of EEG data: it helps the collaborators to "communicate structures and patterns in EEG data by mimicking the patterns with their own voice" (Hermann et al. 2006: 6, cited in Supper 2016: 76).

In such cases, vocalizations, verbal descriptions, and manual notations are not only fast and efficient, but also constitute important epistemological work, similarly to the mathematical writing on blackboards discussed by Barany and MacKenzie (2014). There, "augmentations, annotations, and elisions" on the spot are crucial, and for students the meticulous transcription of those jottings is an embodied entry into thinking along and understanding the math. While such notes may seem to serve as mnemonic devices, the act of note-taking itself is actually at the heart of opening up and grasping knowledge. In fact, the mathematicians who contributed to Barany and MacKenzie's study acknowledged that they rarely returned to their notes, not even those on scrap paper. The notes hardly ever left "the sites in which they were produced" (p. 119). In contrast to Latour's focus on inscriptions as immutable mobiles, Barany and MacKenzie suggest treating the blackboard and scrap-paper scribbles as "immobilized mutables" that themselves constitute creative work (p. 118). In similar ways, vocalizing, verbally describing, and manually transcribing sound transforms sound into sound, sound into words, and sound into images to enact knowledge in close temporal, spatial, and embodied connection with the observation, analysis and explication of the phenomenon concerned.

The situations just described often involve a form of hierarchy, in which one individual—a patient, a student, a motorist, a person on board of a ship equipped with sonar—is dependent on the auditory diagnostic skills of another. This figure—a doctor, a teacher, a mechanic, a sonar operator— has professional jurisdiction about tacit (at times, even secret) knowledge. This both enables and demands a temporary suspension of disbelief by the dependent party. Finally, reading Ingold again helps us understand the special significance of manual notation in situations involving training. Ingold stresses that reading a text is fundamentally different from reading a musical score. Whereas the former is about "taking in" meaning, working "inward," the latter is about "acting out the instructions inscribed in the score," working "outward" (Ingold 2007: 11). In training sessions, the outward direction of sensory instruction through notation or scores is what helps to bring teacher and student onto the same plane.

Conclusions

This chapter has attended particularly to the representation of sound across a variety of disciplines that use sound as a portal for understanding bodies, machines, and other objects of study. It has shown how talking about and transcribing sound in the sciences—whether through verbalizing sound, musical staff notation, or non-automatic graphic notation—both reflected and co-constituted the objects at stake. It has also discussed how and why non-automatic registration of sound survived in some settings and situations despite the rise of mechanical recording.

In late nineteenth-century and early twentieth-century sonar research, medicine, automotive engineering, car mechanics, and ornithology, the use of metaphors, onomatopoeia, and synesthetic translations was very important. In the metaphors, comparisons with sounds familiar from shared soundscapes predominated, ranging from the sounds of human voices and animals to those of natural phenomena, machines, and musical instruments. Two domains, medicine and automotive engineering, witnessed the gradual rise of a dual approach to sound. One aspect was direct sound mapping and the creation of fine-grained classifications and tables of sound that positively referred to particular problems. The other started from a differential diagnosis in which sonic signs merely indicated broad categories of potential problems, the cause of which then needed to be specified through exclusion—that is, negatively. The first strategy was linked to settings that allowed a local and embodied sharing of tacit knowledge; the second thrived in situations where that was not possible.

In most of the sciences and science-related professions discussed in this chapter, musical staff notation played an only marginal role. The exception was ornithology, in which bird sound was initially defined as bird song and music, and could thus follow fields such as music, musicology, ethnomusicology, and folk dance in using musical staff notation in combination with syllabic notation. But in both ornithology and music-related fields, musical staff notation started to receive a more critical press in the first decades of the twentieth century. In music, most of the critics—among whom were many composers, ethnomusicologists, and folklorists—demanded either an expansion of musical staff notation or the introduction of graphic notation in order to account for sounds that could not be captured by conventional musical notation, such as noises, microtones, complex rhythms, and non-stable metrics. In ornithology, arguments that musical staff notation was unnecessarily elitist and that only new forms of notation could capture the rich vocabulary of bird communication—as opposed to bird song—favored graphic notation until sound spectrograms became the standard for visualizing sound.

Despite the seemingly overwhelming victory of mechanical recording in combination with the sound spectrograph, non-automatic and manual recording of sound survived for particular purposes. Some researchers celebrated the analytical strength of embodied musical staff notation, others the efficiency of its simplicity in the field or on the ward, whether for urgent recording when no mechanical recording equipment is available or as a way of evoking sound on-the-spot. Moreover, as musical staff and graphic notation are also instructions for performing sound, its manual forms are highly functional in situations where future expert listeners are being trained in the sonic cultures of their profession, as well as in professional cultures where people have to communicate sound immediately in order to "save the patient."

As well as talking about sound and transcribing sound, this chapter has addressed the contexts in which scientists, doctors, and engineers turned to sound as a source of knowledge in the first place. One final example highlights the contingency of these situations. Discussing the emergence of forerunners to the Geiger-Müller counter in the late 1910s, Axel Volmar (2015) has argued that auditory detection instruments resulted from the desire to measure radiation at a more granular level than was permitted by existing visual techniques. This was done by making the ionization of gases through radiation not only visible, but also audible, with the help of telephones, amplifiers, and later loudspeakers. Telephones responded faster than electrometers with photographic equipment and did not require darkened rooms. The ear also processed such information more quickly than the eye, and discovering the causes of the sounds promised a better understanding of the differences in intensity between α and β radiation.

Volmar shows that without the audio technologies available at the time, it would not have been possible to construct the sounding radiation detectors. But the growing interest in detection by auditory means was also buttressed by widely shared auditory experiences of the World War I front, in which some of the physicists working on the new detection technologies had participated. Forced to analyze the thunder of artillery for their own survival, millions of people had learned to use their ears in acquiring relevant knowledge, and physicists' frequent use of warlike acoustic rhetoric to describe discharge ("atomic drumfire") tapped into these experiences (p. 44). In 1928, Walter Müller, a Ph.D. student of physicist Hans Geiger, constructed the auditory radiation tube that would be given his and Geiger's name. In its mobile commercial version, the Geiger-Müller counter became an iconic tool for localizing sources of radiation, for better or worse. As early as 1929, Müller proudly reported in a letter to his parents that Albert Einstein considered the counter "the most sensitive organ of humanity" (Müller cited in Volmar 2015: 39).

Certainly, the mere availability of audio technologies does not sufficiently explain the turn to sonic skills in this case, nor in the sciences at large. The phonograph was only taken up in the study of birds once the options for amplification of sound matched the ornithologists' growing focus on studying birds in natural settings. In medicine, the stethoscope was not only embraced so enthusiastically by doctors because it enabled them to examine patients without having to rely on their narratives, but also because it permitted male doctors a greater physical distance from female patients. Such historical contexts, however, do not preclude a systematic understanding of why and how scientists, engineers, and physicians listened and listen. That will form the topic of the next chapter.

NOTES

1. A Google Scholar search on August 27, 2013, yielded 12,000 hits for publications about or using the semantic differential method since 2009. The automotive industry is one of the fields employing the method, for example for consumer evaluations of car sound (Cleophas and Bijsterveld 2012). The idea of using pairs of polar terms to study meaning was not Solomon's invention. His system came from research on synesthesia—a phenomenon in which sensations in a particular sensory mode recurrently trigger sensations in another mode—by the American psychologists Theodore F. Karwoski and Henry S. Odbert (1938).
2. For some images, see Booth 2 of our virtual exhibition on sonic skills, at http://exhibition.sonicskills.org/exhibition/booth2/doctors-distance-in-listening/ (last accessed April 21, 2017). There, we also explain the rise of specialized designs for the stethoscope head tailored to either cardiovascular or respiratory sounds.
3. On the International Lung Sounds Association and its Committee on Lung Sound Nomenclature, see http://www.ilsaus.com/pdf/1st_ILSA_1976.pdf, at p. 34 (last accessed August 9, 2016).
4. For the Bobolink transcription, see the second image at http://exhibition. sonicskills.org/exhibition/booth4/notating-bird-song-and-sound/ (last accessed August 14, 2017).
5. See the first image at http://exhibition.sonicskills.org/exhibition/booth4/graphical-notation-the-spectrograph/ (last accessed August 14, 2017).
6. Since the late nineteenth century, this had also be a concern in the world of music itself. Ana Maria Ochoa Gautier has recently explained how the Colombian poet, composer, and musician Diego Fallón aimed to replace musical notation with orthographic notation—notation for the pronunciation of language—to facilitate the making and distribution of music

(Gautier 2014). In Europe and North America, early twentieth-century educators promoted solmization, a system that names notes in terms of their relative rather than absolute position in a key (Whittaker 1924), the use of colored notes (for a discussion, see Wellek 1932), or klavarskribo, a form of graphical notation based on picturing the piano (Pot 1933).

REFERENCES

Abraham, O., & von Hornbostel, E. M. (1909). Vorschläge für die Transkription exotischer Melodien. *Sammelbände der Internationalen Musikgesellschaft*, *11*(1), 1–25.

Barany, M. J., & MacKenzie, D. (2014). Chalk: Materials and Concepts in Mathematics Research. In C. Coopmans, J. Vertesi, M. Lynch, & S. Woolgar (Eds.), *Representation in Scientific Practice Revisited* (pp. 107–129). Cambridge: MIT Press.

Bijsterveld, K. (2007). *Weg van geluid: Hoe de auto een plaats werd om tot rust te komen*. Maastricht: Universiteit Maastricht.

Bijsterveld, K. (2012). Listening to Machines: Industrial Noise, Hearing Loss and the Cultural Meaning of Sound. In J. Sterne (Ed.), *The Sound Studies* (pp. 152–167). New York: Routledge.

Bijsterveld, K., & Krebs, S. (2013). Listening to the Sounding Objects of the Past: The Case of the Car. In K. Franinović & S. Serafin (Eds.), *Sonic Interaction Design* (pp. 3–38). Cambridge: MIT Press.

Bijsterveld, K., Cleophas, E., Krebs, S., & Mom, G. (2014). *Sound and Safe: A History of Listening Behind the Wheel*. Oxford: Oxford University Press.

Brady, Erika. (1999). *A Spiral Way: How the Phonograph Changed Ethnography*. Jackson: University Press of Mississippi.

Bruyninckx, J. (2012). Sound Sterile: Making Scientific Field Recordings in Ornithology. In T. Pinch & K. Bijsterveld (Eds.), *The Oxford Handbook of Sound Studies* (pp. 127–150). Oxford: Oxford University Press.

Bruyninckx, J. (2013). *Sound Science: Recording and Listening in the Biology of Bird Song, 1880–1980* (Ph.D. thesis, Maastricht University).

Bruyninckx, J. (2014). Silent City: Listening to Birds in Urban Nature. In M. Gandy & B. Nilsen (Eds.), *The Acoustic City* (pp. 42–48). Berlin: Jovis.

Bruyninckx, J. (2015). Trading Twitter: Amateur Recorders and Economies of Scientific Exchange at the Cornell Library of Natural Sounds. *Social Studies of Science*, *45*(3), 344–370.

Busoni, F. (1962/1907). Sketch of a New Esthetic of Music. In *Three Classics in The Aesthetic of Music* (pp. 75–102). New York: Dover.

Caravaglios, C. (1935). The Collection and Transcription of Folk-Dances. *Journal of the English Folk Dance and Song Society*, *2*, International Festival Number, 127–135.

Cleophas, E., & Bijsterveld, K. (2012). Selling Sound: Testing, Designing and Marketing Sound in the European Car Industry. In T. Pinch & K. Bijsterveld (Eds.), *The Oxford Handbook of Sound Studies* (pp. 102–124). Oxford: Oxford University Press.

Cowell, H. (1932). Henry Cowell schrijft ons. *Maandblad voor Hedendaagsche Muziek, 1*(12), 90–91.

Douglas, M. (1937). Manx Folk Dances: Their Notation and Revival. *Journal of the English Folk Dance and Song Society, 3*(2), 110–116.

Gallop, R. (1935). Systematization of Motives in the Ceremonial Dance. *Journal of the English Folk Dance and Song Society, 2*, International Festival Number, 79–83.

Gautier, A. M. O. (2014). *Aurality: Listening & Knowledge in Nineteenth-Century Colombia*. Durham, NC: Duke University Press.

Gerard, P. (2002). *Secret Soldiers: The Story of World War II's Heroic Army of Deception*. New York, NY: Dutton.

Gitelman, L. (1999). *Scripts, Grooves, and Writing Machines: Representing Technology in the Edison Era*. Stanford, CA: Stanford University Press.

Goodwin, S. (2010). *Sonic Warfare: Sound, Affect, and the Ecology of Fear*. Cambridge: MIT Press.

Harris, A. (2015). Autophony: Listening to Your Eyes Move. *Somatosphere: Science, Medicine and Anthropology*. Available at http://somatosphere. net/2015/06/autophony-listening-to-your-eyes-move.html. Last accessed August 18, 2017.

Harris, A. (2016). Listening-Touch Affect and the Crafting of Medical Bodies Through Percussion. *Body & Society, 22*(1), 31–61.

Harris, A. & Van Drie, M. (2015). Sharing Sound: Teaching, Learning and Researching Sonic Skills. *Sound Studies: An Interdisciplinary Journal, 1*(1), 98–117.

Hochman, B. (2010). Hearing Lost, Hearing Found: George Washington Cable and the Phono-Ethnographic Ear. *American Literature, 82*(3), 519–551.

Ingold, T. (2007). *Lines: A Brief History*. Milton Park: Routledge.

Karwoski, T. F., & Odbert, H. S. (1938). Color-Music. *Psychological Monographs, 50*, 2.

Kenworthy Schofield, R. (1928). Morris Dances from Field Town. *Journal of the English Folk Dance Society, 2*, 22–28.

Krebs, S. (2012a). "Notschrei eines Automobilisten" oder die Herausbildung des Kfz-Handwerks in Deutschland. *Technikgeschichte, 79*(3), 185–206.

Krebs, S. (2012b). "Sobbing, Whining, Rumbling": Listening to Automobiles as Social Practice. In T. Pinch & K. Bijsterveld (Eds.), *The Oxford Handbook of Sound Studies* (pp. 79–101). Oxford: Oxford University Press.

Krebs, S. (2013). Von Motorkonzerten und aristokratischer Stille: Die Einführung der geschlossenen Automobilkarosserie in Frankreich und

Deutschland, 1919–1939. In R.-J. Gleitsmann & J. Wittmann (Eds.), *Innovationskulturen um das Automobil: Von gestern bis morgen* (pp. 77–99). Königswinter: Heel.

Krebs, S. (2014a). "Dial Gauge versus Sense 1-0": German Car Mechanics and the Introduction of New Diagnostic Equipment, 1950–1980. *Technology and Culture, 55*(2), 354–389.

Krebs, S. (2014b). Diagnose nach Gehör? Die Aushandlung neuer Wissensformen in der Kfz-Diagnose (1950–1980). *Ferrum: Wissensformen der Technik, 86,* 79–88.

Lachmund, J. (1999). Making Sense of Sound: Auscultation and Lung Sound Codification in Nineteenth-Century French and German Medicine. *Science, Technology and Human Values, 24*(4), 419–450.

Marsden, W. J. M. (1927). Some Observations on Bird Music. *Music & Letters, 8*(3), 339–344.

Martin, M., & Fangerau, H. (2011). Töne sehen? Zur Visualisierung akustischer Phänomene in der Herzdiagnostik. *NTM Zeitschrift fur Geschichte der Wissenschaften, Technik und Medizin, 19*(3), 299–327.

Maslen, S. (2015). Researching the Senses as Knowledge: A Case Study of Learning to Hear Medically. *The Senses & Society, 10*(1), 52–70.

Mundy, R. (2009). Birdsong and the Image of Evolution. *Society and Animals, 17*(3), 206–223.

Mundy, R. (2010) *Nature's Music: Birds, Beasts, and Evolutionary Listening in the Twentieth Century* (Ph.D. thesis, New York University).

Osgood, C. E., Suci, G. J., & Tannenbaum, P. H. (1957). *The Measurement of Meaning.* Urbana: University of Illinois Press.

Pot, C. (1933). Klavarskribo: Proeve van een vereenvoudiging van ons notenschrift (Slot). *Maandblad voor Hedendaagsche Muziek, 2*(4), 129–130.

Reichert, S., Gass, R., Brandt, C., & Andrès, E. (2008). Analysis of Respiratory Sounds: State of the Art. *Clinical Medical Insights: Circulatory, Respiratory and Pulmonary Medicine, 2,* 45–58.

Rice, T. (2013). *Hearing and the Hospital: Sound, Listening, Knowledge and Experience.* Canon Pyon: Sean Kingston Publishing.

Rieger, S. (2009). *Schall und Rauch: Eine Mediengeschichte der Kurve.* Frankfurt am Main: Suhrkamp Verlag.

Ross, C. D. (2004). Sight, Sound, and Tactics in the American Civil War. In M. Smith (Ed.), *Hearing History: A Reader* (pp. 267–278). Athens: University of Georgia Press.

Russolo, L. (1986/1916). *The Art of Noises.* New York: Pendragon Press.

Schwartz, H. (2012). Inner and Outer Sancta: Earplugs and Hospitals. In T. Pinch & K. Bijsterveld (Eds.), *The Oxford Handbook of Sound Studies* (pp. 273–297). Oxford: Oxford University Press.

Solomon, L. N. (1954). *A Factorial Study of the Meaning of Complex Auditory Stimuli (Passive Sonar Sounds)*. Urbana: University of Illinois.

Sovijärvi, A. R. A., Dalmasso, F., Vanderschoot, J., Malmberg, L. P., Righini, G., & Stoneman, S. A. T. (2000). Definition of Terms for Applications of Respiratory Sounds. *European Respiratory Review, 10*(77), 597–610.

Stockmann, D. (1979). Die Transkription in der Musikethnologie: Geschichte, Probleme, Methoden. *Acta Musicologica, 51*(Fasc. 2), 204–245.

Supper, A. (2012). *Lobbying for the Ear: The Public Fascination with and Academic Legitimacy of the Sonification of Scientific Data* (Ph.D. thesis, Maastricht University).

Supper, A. (2016). Lobbying for the Ear, Listening with the Whole Body: The (Anti-)Visual Culture of Sonification. *Sound Studies: An Interdisciplinary Journal, 2*(1), 69–80.

Van Drie, M. (2013). Training the Auscultative Ear: Medical Textbooks and Teaching Tapes (1950–2010). *The Senses and Society, 8*(2), 165–191.

Volmar, A. (2012). *Klang als Medium wissenschaftlicher Erkenntnis: Eine Geschichte der auditiven Kultur der Naturwissenschaften seit 1800* (Ph.D. thesis, Universität Siegen).

Volmar, A. (2013). Listening to the Cold War: The Nuclear Test Ban Negotiations, Seismology, and Pyschoacoustics, 1958–1963. In A. Hui, J. Kursell, & M. Jackson (Eds.), Music, Sound and the Laboratory from 1750–1980. *Osiris, 28*, 80–102.

Volmar, A. (2014). In Storms of Steel: The Soundscape of World War I and Its Impact on Auditory Media Culture During the Weimar Period. In D. Morat (Ed.), *Sounds of Modern History: Auditory Cultures in 19th- and 20th-Century Europe* (pp. 227–255). New York, NY: Berghahn.

Volmar, A. (2015). Ein "Trommelfeuer von akustischen Signalen": Zur auditiven Produktion von Wissen in der Geschichte der Strahlenmessung. *Technikgeschichte, 82*(1), 27–46.

Wellek, A. (1932). Die Entwicklung unserer Notenschrift aus dem Tönesehen. *Acta Musicologica, 4*(3), 114–123.

Whittaker, W. G. (1924). The Claims of Tonic Solfa I. *Music & Letters, 5*(4), 313–321.

Zwikker, C. (1934). De oorzaken van het geluid bij automobielen. In Anonymous, *Verslag van het 'Anti-lawaai Congres,' georganiseerd te Delft, op 8 november 1934 door de Koninklijke Nederlandsche Automobile Club in samenwerking met de Geluidstichting* (pp. 70–77). Delft: KNAC/Geluidstichting.

CHAPTER 3

Modes of Listening:
Why, How and to What?

Abstract This chapter presents a typology of the modes of listening employed across science, medicine, and engineering. It distinguishes between three *purposes* of listening and three *ways* of listening in the sciences. The three purposes discussed are diagnostic, monitory, and exploratory listening; the three ways are analytic, synthetic, and interactive listening. Using ample examples, this chapter illustrates the six modes of listening and the virtuoso mode-switching of scientists and other experts. It reflects on the incidence of specific combinations of purposes and ways of listening, and asks how these listening modes interact with the third dimension of listening: listening to *what*.

Keywords Diagnostic listening · Monitory listening · Exploratory listening · Analytic listening · Synthetic listening · Interactive listening

---

Most of this chapter has been published in open access as: Supper, A., & Bijsterveld, K. (2015). Sounds Convincing: Modes of Listening and Sonic Skills in Knowledge Making. *Interdisciplinary Science Reviews, 40*(2), 124–144. Four of the six listening modes discussed below have been first introduced in Bijsterveld's 2009 grant proposal (see footnote 8 in Chapter 1), and in Pinch and Bijsterveld (2012: 14).

K. Bijsterveld, *Sonic Skills*,
https://doi.org/10.1057/978-1-137-59829-5_3

61

Introduction

In 1917, the American composer and educator Sophie Gibling published an essay on "Types of Musical Listening." The kind of typology it presented held much sway with cultural commentators and musical critics in the early twentieth century: it distinguished between imperfect and ideal listeners. Among several types of imperfect listeners was the listener "who is no listener at all, who passively sits through a concert, intellectually contributing nothing; waiting, like a cabbage or a stone, for something to happen to him. He hears without listening" (Gibling 1917: 386). The ideal listener, in contrast, prepared himself for concerts emotionally and intellectually, listened past imperfections in specific performances to appreciate the beauty of the composition, and was ready to merge completely with the music, becoming a "purely abstract spirit" (p. 389). To Gibling, listening well was a matter of the "quality of a man's personality" (p. 388).

In the century following the publication of Gibling's essay in the *Musical Quarterly*, many more typologies of modes of listening have appeared, spanning fields as diverse as cultural studies, musicology, media studies, communication studies, and psychoacoustics. Like Gibling, some authors display strong preferences for particular modes, while others question the value of normative judgments (Stockfelt 1997; Subotnik 1991). Alongside those concerned with musical listening, authors have offered taxonomies of listening for domains such as radio broadcasting (Douglas 1999; Goodman 2010), film sound (Chion 2005/1990), and everyday environments (Truax 2001/1984).

Based on work published by Alexandra Supper and myself in 2015, this largely co-written chapter extends the discussion of taxonomies of listening to yet another set of empirical domains: science, medicine, and engineering. We developed a two-dimensional taxonomy of the listening practices in the sciences, one that takes into account both *purposes* of listening (the why) and *ways* of listening (the how). A particular strength of this taxonomy is that it allows us to show how practitioners shift between different modes of listening—an ability that is at least as important for knowledge making as is competence in using any one given mode of listening. Our aim is not to isolate listening from other skills, but to argue that our understanding of knowledge dynamics can be substantially deepened by attending to the ways in which listening modes inform the use of sonic skills in knowledge-making processes.

As explained in Chapter 1, our use of the term *mode* does not imply an exclusive focus on the cognitivist dimension. We regard the modes of listening as being linked to particular bodily practices and embedded in a broader set of sonic skills. Sonic skills, in our approach, include both listening skills and the techniques that doctors, engineers, and scientists employ for making, recording, storing, and retrieving sound.

The goal of this chapter is thus to show how listening modes, and the ways in which they feed into sonic skills, cast light on processes of knowledge production in science, engineering, and medicine. We first explain the origins of our typology of listening modes, which has drawn inspiration both from existing scholarly work in sound studies and from actors' categories. We then outline our typology in more detail. In the final two sections, we outline how an analysis of listening modes can usefully be integrated into a study of broader sonic skills, and substantiate the relevance of both notions—listening modes and sonic skills—for understanding processes of knowledge production.

EXISTING TAXONOMIES OF LISTENING: ANALYTIC AND ACTORS' CATEGORIES

Thinking about listening modes has a long tradition both in the academic field of sound studies and among practitioners who work with sound. Perhaps the best-known typology of modes of listening—although it may be more accurately characterized as a typology of *listeners*—is the one developed by Theodor Adorno (1977/1962), which distinguishes between figures such as experts, good listeners, culture consumers, emotional listeners, and entertainment listeners. Adorno makes no secret of his preference for structural listening, a mode commonly displayed by experts, nor of his contempt for practices such as entertainment listening. In this preference, Adorno's typology echoes the concerns of many music theorists and critics since the middle of the nineteenth century, beginning with the Viennese music critic Eduard Hanslick, who offered an opinionated taxonomy to distinguish "poor listening practices from the true method of listening, *aesthetic listening*" (Hui 2013: 34). Similar taxonomies accumulated in the course of the nineteenth and early twentieth century, usually taking attentive, absorbed listening as their gold standard.[1] Anxieties about proper listening modes were not limited to the world of classical music; they also took hold in the domain of radio broadcasting. Historians of broadcasting have traced public debates about proper modes of radio listening back to the 1930s, when many commentators warned of the social, political, and psychological dangers of *distracted listening* (Goodman 2010) and tried to persuade "the listener to 'incline their ear' not only in the right direction (towards beautiful, honest and reputable things), but also in the right way (selectively, attentively and with appropriate discrimination)" (Lacey 2013: 183).

In the course of the twentieth century these typologies, with their outspoken normative preferences, increasingly came under fire or were put into historical perspective. Fervent critiques of Adorno's categorization scheme and his advocacy for structural listening were expressed within musicology (Subotnik 1991, 1996; Stockfelt 1997), and inspired an abundance of work proposing relativist and postmodern alternatives (Dell'Antonio 2004). Music theorist Ola Stockfelt (1997), for instance, argued that different modes of listening are appropriate for and indeed demanded by different genres—Adorno's preferred mode, structural listening, being adequate only for a very specific type of Western art music. Modes of listening should therefore be judged in terms of adequacy to a genre: adopting an adequate mode means being able "to listen for what is relevant to the genre" (Stockfelt 1997: 137). With this assertion, Stockfelt moves away from treating modes of listening as personal characteristics (linked to particular character traits or socioeconomic factors), and instead treats them as part of a repertoire from which individuals can choose. Moving between different modes is not only possible, but common.

This possibility of *shifting* between modes of listening (often akin to shifting between levels of attention) has been an important element of many recent typologies. For instance, in his book *Acoustic Communication*, Barry Truax (2001/1984) distinguishes three modes of listening, each characterized by a different level of attention. The first kind, *background listening*, is entirely passive, listening that is not directed at achieving any practical purpose. By comparison, *listening-in-readiness* (such as the practice of recognizing a vehicle by its sound) is more active, while *listening-in-search* (as exemplified by a ship captain whistling and using the echo for orientation) is more active still. Truax's research is embedded in a normative concern about noise pollution in modern society, as he worries that the skills of listening-in-readiness and listening-in-search are rapidly dwindling due to the noise of modern technology.

In her book *Listening In*, historian of technology Susan J. Douglas (1999) offers an "archaeology of radio listening" that goes well beyond a distinction between concentrated and distracted listening practices. Among the many listening modes she mentions are *linguistic, musical, informational, exploratory, story, advertisement*, and *fidelity listening* (1999: 33–35). Some of Douglas's categories describe listening in terms of what someone is listening *to* (stories or ads, for example), some in terms of what they listen *for* (such as information or sound quality), and some in terms of *how* they listen cognitively. Examples of these cognitive ways of listening are *dimensional listening*, in which the listener imagines spaces, and *associational listening*, in which networks of memories are triggered through sound.

In this chapter, we follow in Douglas's footsteps to develop a typology that operates on more than one dimension, but we do so in a way that explicitly asks how these dimensions relate to each other. Like Douglas and other scholars, we start from the idea that listeners have a repertoire of listening modes available between which they can shift. We go one step further, arguing not only that shifting between different modes of listening is possible and common, but that the capability of shifting is itself an essential skill in the knowledge-making practices of many scientists, engineers, and doctors.

Before delving into our own typology of listening modes in the sciences, we should briefly acknowledge that it was inspired not only by scholarly work in sound studies, but also by the discourses of the actors that we studied. That is to say, some of our categories are based on actors' categories. For instance, the distinction between "monitory listening" (listening to monitor whether everything is working well) and "diagnostic listening" (listening to diagnose the specific source or cause of a problem), which we will explore in more detail in the next section, is present in the discourse of car mechanics themselves. It played an important role in the formalization and professionalization of the trade of German car mechanics during the 1930s, as discussed by Stefan Krebs (2012a, b, c) and in this essay's previous chapter. To convince car drivers to entrust their faulty cars to the new profession of certified car mechanics, the mechanics needed to ensure that car drivers would trust their own ears enough to know when to bring their car to the garage, but not enough to try to fix it themselves. Once it was established that there was a problem, the task of diagnosing and fixing that problem would be left to the mechanics. In their effort to gain exclusive jurisdiction over the ability to repair cars, the mechanics demarcated their skill of professional diagnostic listening from the monitory listening skills of drivers. The distinction between two listening modes, in other words, helped the mechanics in their quest for cultural authority.

Explicit references to taxonomies of listening modes are even more widespread in the scientific community dedicated to sonification research, which aims to systematically explore the use of sound to represent data and convey information. Indeed, the distinction between different modes of listening has been a stable feature of the sonification literature, from some of the founding texts of the community (Gaver 1989; Williams 1994) to much more recent contributions (Vickers 2012; Grond and Hermann 2014). An example is William Gaver's (1989) work on the SonicFinder, an early attempt to use auditory displays as part of a computer interface, which was a forerunner of now-ubiquitous sounds such as those announcing new emails or accompanying the moving of a file into a digital trashcan. The SonicFinder builds upon a fundamental distinction between two modes of listening: everyday listening and musical listening. The former is directed at identifying the sources of a sound, the latter at its formal characteristics, such as pitch or timbre. Gaver's displays are designed to exploit everyday listening in particular: "If sounds are to be used in the interface, they should be used much as they are in our everyday lives …. We do not hear the pitch of closing doors; instead we are more likely to hear their size, the materials from which they are made, and the force used to shut them" (Gaver 1989: 72ff). This distinction still resonates in sonification research today, and—although many sonifications do demand a certain degree of musical listening skill, as pitch is a widely used parameter in sonification designs—so does the emphasis on everyday listening skills (Hermann 2011). The sonification community struggles with the fact that many potential end users of sonification (specialists in the scientific domains from which data are translated into sound) are reluctant to use sonification because they distrust their own ears.[2] In that situation, references to everyday listening can reduce the fear of listening, and thus potentially help to convince people of the benefits of sonification.

Similarly, scholars in the sonification community frequently make a distinction between *synthetic listening* (listening to sounds holistically) and *analytic listening* (focusing on specific elements of the sound), which has been appropriated from literature in auditory perception research (Hermann 2002; Worrall 2009; Walker and Nees 2011; Williams 1994).

This distinction also turned out to be useful for our own typology, which adopts the categories of analytic and synthetic listening—along with an additional category of interactive listening—as two fundamentally different ways of listening.

Our typology thus makes use of existing categories employed by the actors that we study, but only as a partial inspiration, in tandem with secondary literature.[3] The most important contribution of our typology above and beyond the existing actors' categories is that it addresses different modes of listening in two dimensions, and places them in a broader context of the sonic skills that are involved in knowledge production.

PURPOSES OF LISTENING: WHY SCIENTISTS, ENGINEERS, AND PHYSICIANS LISTEN

Our proposed typology takes into account both the purposes for which scientists, engineers, and physicians listen and the ways in which they do so. In this section, we set out three purposes of listening—monitory, diagnostic, and exploratory—before adding three different ways of listening—synthetic, analytic, and interactive—to arrive at nine possible combinations of modes.

Monitory listening refers to checking for possible malfunctions—for instance, when car drivers pay attention to "the rhythmic and silent run of the engine" and "the regular humming of the gearbox or chain drive" (Küster 1919, cited in Bijsterveld and Krebs 2013: 20). Monitory listening is also employed in the scientific laboratory and field by researchers checking the proper running of their equipment (Bruyninckx 2013; Mody 2005), and in the hospital by doctors and nurses monitoring the vital signs of patients. Monitory listening usually accompanies other tasks and activities, often unrelated to sound—whether driving a car, operating a microscope, or performing surgery on a patient. The fact that the sound can be perceived in the background while focusing on other tasks, but that sudden and unexpected changes in the sound nonetheless immediately draw the listener's attention, is of great benefit here. Similarly, the ability to "monitor multiple processes simultaneously" (Dayé and de Campo 2006: 350) has been considered an advantage of sonification over graphic displays. Consequently, sonifications developed specifically for monitoring have been an important area of sonification research in recent years, ranging from applications developed for the medical field to those monitoring web server activity. It has even been claimed that the monitoring of information in the background while users attend to another task is where sonification and auditory display really "come into their own" (Vickers 2011: 456).

Whereas monitory listening is concerned with establishing *whether* something is wrong, diagnostic listening is about pinpointing *what* precisely is wrong. The quintessential example of diagnostic listening is that of physicians using their stethoscope during physical examinations to distinguish the "normal" sounds of a healthy body from the "abnormal" sounds of a sick one and to diagnose specific diseases based on those sounds. This skill is not limited to the medical field. In fact, practitioners from other domains frequently reference the listening practices of doctors to explain their own: "If the physician cannot make his diagnosis by the appearance of the patient, he will take his stethoscope and listen to the patient's body. This is how you ought to proceed with the car engine as well," proposed one car mechanics handbook (Hessler 1926, cited in Krebs 2012c: 83). When physicist Cornelis Zwikker (1934) discussed sixteen car sounds and their causes, he described "knocking" as a symptom of an "advanced-stage disease" (p. 75). As Stefan Krebs and Melissa Van Drie (2014) have shown in their comparative work on doctors and mechanics, such use of medical metaphors and pictures staging mechanics as "car doctors" became widespread from the 1920s onward. As we have seen, medical techniques such as differential diagnosis became models for automotive engineering, just as the trust invested in the medical profession became a model for the role of the car mechanic (pp. 95–97). In ornithology, the skill of diagnostic listening was, and still is, considered essential for the correct identification of species, but also for ensuring adequate recording quality—examples will follow in Chapter 4. And in sonification, diagnostic listening plays an important role in quality control, as errors in sonification design are often picked up by listening.

Exploratory listening, thirdly, refers to listening out for new phenomena. The notion was developed by Douglas (1999) for the practice of radio hams trying to discover distant stations, but it also plays a part in the listening practices of scientists. Narratives of field observation in ornithology, for instance, often feature ornithologists letting themselves be guided through the woods by ear, always listening out for rare, exotic, or appealing bird songs, such as in this account by the naturalist J. Schafer:

> While going through a thicket of hazel brush, briars and vines, a bird was heard singing so softly that it was some time before I could locate the

exact place where the song came from. After listening a short time I recognized the song to be that of a Catbird, but to make sure of the identity
of the singer, it was driven from its hiding place. (Schafer 1916, cited in
Bruyninckx 2013: 36)

Although sonification is usually a more mundane activity, taking place
with headphones in front of a computer screen, the exploratory listening of sonification researchers, too, can become entangled with romantic
narratives of adventurous scientists making chance discoveries thanks to
their dedicated attention to their sonic environment—as in this description of Robert Alexander's solar wind sonifications:

> Alexander typically compresses 44,100 data points into a second of sound,
> the sampling rate of a compact disc.
>
> Then, he puts on his headphones.
>
> On that particular day he found a hum everywhere in the data.
> "I thought I was hearing noise," he recalls.
>
> But it was more than that. The hum had a frequency of 137.5 hertz which
> would correspond to about 26 days in the original data. That would be
> the time taken for a particular feature on the sun to swing back around. In
> other words, he could lock on a feature and listen in.
>
> Alexander realized what he was hearing and messaged a colleague. "The
> frequency I'm listening to is the rotation speed of the sun. I don't think
> anyone's ever done this."[4]

In a recent video documentary for *Vice Magazine*, Alexander explains the
exploratory nature of his listening: "I was digging through, you know, 20
or 30 different data parameters and listening to them all, and I realized
that if I listened to carbon, that I could hear a very strong harmonic presence."[5] In the astrophysical research group he was working with, carbon
had not previously been mentioned as relevant to the study of solar wind;
instead, different types of solar wind had been distinguished by measuring
oxygen charge states. It was through listening to different sonic realiza-
tions of the same dataset that Alexander's research group became aware of
the potential of carbon as a more reliable indicator for solar wind activity.
The results were written up and published in the *Astrophysical Journal*—
with a brief mention of the sonification process that had led to the discovery, but none of the romantic flourishes of a lone researcher making a
chance discovery when donning his headphones (Landi et al. 2012).

WAYS OF LISTENING: HOW SCIENTISTS, ENGINEERS, AND PHYSICIANS LISTEN

The three modes of listening discussed so far were concerned with the purposes for which scientists, engineers and physicians listen; in the following, we introduce three modes that describe the ways in which they do so: synthetically, analytically, or interactively. These three modes do not exclude the purpose-related modes; rather, any given listening practice of a scientist, engineer, or physician can always be characterized both in terms of its purpose and in terms of its manner. A car driver listening to the roar of the engine while driving, for instance, engages in both monitory and synthetic listening.

The term *synthetic listening* comes from literature in auditory perception research, and has become a mainstay of sonification literature. Its meaning is usually defined in opposition to another category in our typology, *analytic listening*. For instance, in the first book publication on sonification, the terms were defined as follows:

> **Synthetic** perception takes place when the information presented is interpreted as generally as possible; for example, hearing a room full of voices or listening to the overall effect of a piece of music. **Analytic** perception takes place when the information is used to identify the components of the scene to finer levels; for instance, listening to a particular utterance in the crowded room or tracking one instrument in an orchestral piece or identifying the components of a particular musical chord. (Williams 1994: 98)

This definition of synthetic listening and analytic listening is also applied in Albert Bregman's (1994) influential work on auditory perception, and still resonates in sonification discourse today (Hermann 2002; Worrall 2009; Walker and Nees 2011). For sonification, both synthetic and analytic listening play a role—both the ability to perceive complex auditory events as a whole, and the ability to break the whole down into its component pieces and single out particular streams of sound for attention. In addition, the capacity to switch between these different modes is considered an important asset for the use of sonifications. The experience of attending a concert is often used as an example both of how people can perceive a piece of music as a whole and of how they analytically attend to specific instruments: "in a concert hall we can hear a symphony orchestra as a whole. We can also tune in our focus and attend to individual musical instruments or even the couple who is whispering in the next row" (Hermann et al. 2011: 3).

When medical students learn to use their stethoscopes, they are first learning the skills of analytic listening: navigating an initially confusing world of sound by differentiating the sounds of the patients' bodies from the sound produced by the tool itself and the sound of their own body. However, the skill of synthetic listening is equally important for the practices of scientists, engineers, and mechanics. Recall the 1965 article for apprentices in mechanics cited in the previous chapter, which claimed that diagnosis only starts when one complex sound is disaggregated into many single sounds. Frequently, successful use of sonic skills involves the combination of analytic and synthetic listening at different stages of the process of knowledge production. The quick identification of a bird in the field, for example, often involves synthetic listening, as ornithologists listen for general features and recognize the bird "more by the quality or style, or both, of its utterance than by the number and succession of its notes" (Summers 1916: 79). Once that quick identification has been made, the ornithologist may listen analytically to rule out confusion with similar-sounding species, or to notate specific elements of the sound.

Although synthetic and analytic listening are usually defined as opposites, they have one important aspect in common: both modes assume that the sound source itself is stable or unfolds according to its own dynamic rules. In many instances where scientists, engineers, and physicians listen, however, they actually intervene into the sounds while listening. We therefore distinguish an additional way of listening, that of *interactive listening*. If synthetic listening means hearing the whole orchestra and analytic listening means focusing on a particular stream of sound (perhaps the second oboe), interactive listening means that the listener decides to replace the second oboe with a didgeridoo halfway through in order to better grasp the dynamics of the piece. Scientists, physicians, and engineers often engage in such interactive listening in order to find out more about their subjects. Ornithologists may interact with the birds that they study by deliberately exposing them to specific sounds—such as recordings of birdsongs or traffic noise—in order to elicit a response (Bruyninckx 2013: 94ff). Car mechanics, too, often engage in interactive listening, for instance when listening for changes in the sound of the engine while changing gears; and so do car drivers, when they pay attention to the sounds of the car in deciding when to shift gears. Interactive listening to car engines can thus serve both monitory purposes (for drivers) and diagnostic purposes (for mechanics).

Interactive listening is also common in sonification research, where it is used mainly for diagnostic or exploratory purposes.[6] As our ethnographic research has shown, diagnostic interactive listening is especially common during the design process: errors in the sonification design (or even in the underlying dataset) often express themselves as discrepancies between expected and actual sounds, and can be corrected by alternately adjusting settings and listening to the results until the expectations and the outcome are aligned. Exploratory interactive listening has also become increasingly popular in sonification research, a trend attested by the growth of a whole subfield dedicated to "interactive sonification" (Hermann and Hunt 2005, 2011). In interactive sonifications, users can "change selections quickly and easily to gain multiple auditory viewpoints" (Flowers 2005: 4), which are intended to give a better understanding of the data, especially for exploratory tasks.

At first glance, the categories of synthetic, analytic, and interactive listening may seem similar to Barry Truax's (2001/1984) distinction between background listening, listening-in-readiness, and listening-in-search. Truax's example of listening-in-search, in which a ship captain whistles and uses the resulting echo for navigation, could be considered an instance of interactive listening. However, in Truax's classification, the three modes are distinguished by the different degrees of active attention paid by the listener, whereas in our scheme it is a matter not of different degrees, but of different *kinds* of attention. And although the categories of synthetic and analytic listening emerged from psychoacoustic research, our approach does not assume that these listening skills are limited to the mind only. Rather, they are enmeshed with particular bodily strategies—examples might be a doctor percussing a patient's chest, an ornithologist cupping his hands around his ears and "rotating slowly like an aural CCTV camera" (Lorimer 2008: 391) while listening out for a particular bird, or a sonification researcher convulsing in pain when enduring unpredictable and piercing sounds in the attempt to identify errors in the sound-generating computer code.

Distinguishing modes of listening on two dimensions can give us a multilayered and nuanced appreciation of the listening practices involved in scientific research, medical work, and engineering. Looking at only one dimension in isolation would give us a very partial understanding of these practices. For instance, if we ask only *why* scientists, engineers,

Table 3.1 Overview of listening modes

Why / How	Synthetic listening	Analytic listening	Interactive listening
Monitory listening	Attending to overall features of sound for the purpose of monitoring	Attending to specific characteristics of sound for the purposes of monitoring	Interacting with a sound source for the purposes of monitoring
Diagnostic listening	Using a (quick) overall impression of a sound for the purposes of diagnosis	Attending to specific characteristics of a sound for the purposes of diagnosis	Interacting with a sound source for the purposes of diagnosis
Exploratory listening	Listening out for general impressions for the purposes of exploration	Attending to specific features of sound for the purposes of exploration	Interacting with the sources of a sound for the purposes of exploration

and physicians listen, we may miss the particular bodily and cognitive skills that their listening entails. Likewise, there is not one single technique for monitory listening—listening for the purposes of monitoring may involve either focusing on general patterns of sound (as car drivers do when listening out for the auditory feedback of their car engines and surroundings) or focusing on particular elements of that sound (as physicians do on their daily ward round when checking whether a symptom discovered during yesterday's diagnosis has cleared up), or even interacting with the source of that sound (as ornithologists do when playing a sound recording to a bird to elicit a reaction). On the other hand, addressing the dimension of *how* we listen in isolation, without taking into account the purposes, risks losing sight of why those sonic skills matter in the first place. The listening modes could then be taken as ends in themselves, rather than as analytical tools telling us something about how scientists, engineers, and physicians use their bodies and senses for particular ends that in and of themselves may not have anything to do with sound.

Our graphic representation of all possible combinations of purposes and ways of listening (Table 3.1) is not, however, intended to suggest that all the listening modes mentioned are equally predominant in the sciences. We noted above that listening practices in the sciences can always be described in terms of both purposes and ways of listening. We have also given empirical examples of all the potential combinations of purposes and ways of listening as listed in the cells of Table 3.1. Our cases studies have thus demonstrated the existence of all these options—but that does not mean all modes or all possible combinations have the same *incidence* in the knowledge practices of scientists, engineers, and physicians.

We have come across numerous examples of analytic listening for the purpose of diagnosis, and synthetic listening for the purpose of monitoring, but our case studies offered only a few illustrations of interactive listening for the purpose of monitoring. Apart from ornithologists playing recorded bird sound to make birds respond and betray their presence at a particular field site, we have only one other example. In intensive care units, all kinds of machines help the medical staff to monitor the physical condition of patients. These instruments' alarms, Anna Harris has explained, can be tweaked manually, meaning that the staff can define the parameters that will set off the alarms. Nursing students are told "to set their alarms 'wide,' so that they are alerted to even the slightest deviance in heart rate or blood pressure." Experienced nurses, in contrast, may tighten the alarms to reduce the number of alerts, "as they have the expertise which enables them to monitor a patient without the continual sounding of alarms" (2015: n.p.).

SONIC SKILLS: VIRTUOSITY IN SHIFTING MODES AND HANDLING TOOLS

Although Table 3.1 might initially give an impression of stagnancy and rigidity, its strength lies in providing a stable reference point for the dynamic listening practices that we found. The professional status of some practitioners—such as doctors or car mechanics—is intimately connected with their recognition as expert diagnostic listeners. Yet, we argue in this section, it is often the ability to shift between different modes of listening, rather than specialization in one particular mode, that expresses the virtuosity of their sonic skills and helps them to underpin their knowledge claims. Furthermore, this ability to shift is closely linked to the handling of tools and to broader sonic skills that go beyond the modes of listening themselves.

In the everyday knowledge practices of the experts we studied, the different listening modes often build upon each other. For instance, it is important for the successful work of car mechanics—for which diagnostic listening is essential—that car drivers recognize the need to bring their car to the garage, for which monitory listening is crucial. In some instances, a shift in the purpose of listening goes hand in hand with a shift in the way of listening. The solar wind sonifications described above were, at least at first, an example of synthetic exploratory listening, as Robert Alexander somewhat randomly listened for general patterns in the data in the hope that something of interest would jump out at him. Once he noticed harmonic presences in the charge states of carbon, however, a shift seems to have occurred towards listening to a particular element in order to diagnose the dynamics of solar wind activity. That involved focusing his attention on one specific aspect of the sound. In other words, there was a shift not only in the purpose of listening (from exploratory to diagnostic), but also in the way of listening (from synthetic to analytic).

In the last two examples, diagnostic listening followed monitory or exploratory listening, while analytic listening followed synthetic listening. It might be tempting to conclude that the modes of listening occur in a fixed order, with diagnostic and analytic listening as the natural culmination and end-point. This is not the case, however. An instance from ethnographic research in the hospital is one illustration that monitory listening does not always precede diagnostic listening. During the initial examination after a patient is admitted to hospital, doctors usually engage in diagnostic listening. But when performing subsequent check-ups during their daily rounds, they are more likely to perform monitory listening, checking whether specific symptoms detected during earlier diagnoses persist or have improved.

In fact, the different modes often occur in constant back-and-forth shifts, with the listener repeatedly zooming in and out. Many sonification designs—especially those made for the purposes of data exploration—are deliberately built to facilitate rapid shifts between different modes of attention. In Thomas Hermann's (2002) dissertation on sonifications for exploratory data analysis, for instance, the listener is described as being engaged in different modes of listening, and many of the proposed sonification designs feature multiple streams of data that can be listened to simultaneously or separately. They seem to be designed for a listener who may at times synthetically listen to several streams of sound simultaneously, and at other times analytically hone in on particularly promising specific sound streams. That listening is interpolated, of course, with the occasional intervention into the sound source itself. Constantly switching between analytic, synthetic, and interactive modes of listening is thus facilitated and intended by the sonification design.

This example flags up the connection between the ability to shift between modes of listening and the availability of particular *tools*. The introduction of a novel type of stethoscope in German automotive engineering during the interwar period would be another, earlier example of how tools can effectively enable the process of mode-shifting. Whereas listening rods and traditional stethoscopes had been used to listen to one component of the car engine, the new stethoscope had two sensors. These sensors, the Tektoskop and the Tektophon, enabled mechanics to listen to two differently located components of the engine simultaneously and make a detailed comparison of the sounds they heard (Anonymous 1929b, cited in Bijsterveld et al. 2014: 80). Such a construction afforded shifts between synthetic and analytic listening, as it brought together distinct sounds in one listening "frame" while keeping alive the option of alternation in auditory focus.

Similarly important for the skill of mode-shifting was the rise of sound recording technologies in ornithology. The phonograph, sound camera, gramophone, and tape recorder enabled ornithologists to record the sounds of birds in ways that many considered more accurate, less dependent on individual listening capacities, and therefore more "objective" than the notation of sound in traditional onomatopoeic terms, musical staff, or graphic systems (Bruyninckx 2013: 63). In addition, these tools made it possible for ornithologists to repeat their listening exercises as often as they wanted, or even to slow down the recording, thus improving precision in notation after the field trip (Bruyninckx 2013: 49). This enabled cycles of analytic listening, with repeated listening allowing ornithologists to focus on different components of bird sound across different listening sessions.

Slowing down gramophone playback was useful not just for those ornithologists who notated bird sound manually, but also for those who favored automated visualizations. For instance, the British ethologist and ornithologist William Homan Thorpe, a strong advocate of the spectrograph (to recall: an instrument that visualizes the frequency and intensity of sound over time), acknowledged that the analysis of sound spectrograms was best accompanied by listening to the sound recordings and, in particular, by doing so at reduced speed. According to Thorpe, slowing down gramophones from 78 to 28 cycles made the complexities and varieties of bird sound—such as high frequencies and rapid sequences—more easily accessible to the human listener and enabled a focus on different components of the sound than those that prevailed at 78 cycles. Thorpe cautioned, however, that recordings played back at reduced speed would "at first hearing have no apparent resemblance to the original" (Thorpe 1958: 542). Despite the analytic value of listening at reduced speed, synthetic listening was often best accomplished at the original speed. An important function of the "infinitely variable speed turntable" favored by Thorpe (1958: 542) was precisely that it enabled scientists to quickly and easily switch not just between different speeds of playback, but also between different modes of listening. The availability and use of particular recording and playback tools thus affected the options for modes of listening, and therefore the character of mode-shifting, which in turn fed into the knowledge claims that were formulated.

Tools can open up particular modes of listening and particular means of shifting between modes, but they may also enhance the epistemological status of listening practices in science, medicine, and engineering. As Tom Rice (2008, 2010) and Melissa Van Drie (2013) have shown, the stethoscope is an important signifier, a visual icon, of the doctor's expertise and jurisdiction. Even though the practical importance of auscultation for medical diagnosis has declined over the years, the stethoscope has retained its symbolic function. Indeed, its symbolic sway reaches beyond the confines of the medical field, as other professional groups have also appealed to the stethoscope's symbolic authority. In the automotive industry, for instance, engineers and mechanics have often been portrayed in white coats and using a stethoscope on a car engine—explicitly alluding to the image of a doctor using a stethoscope to examine a patient (Krebs and Van Drie 2014). Tools may thus function as symbolic capital underlining the epistemological authority of their users.

It is important to note, once again, that the sonic skills involved in knowledge-making practices are not a matter of listening alone. The examples of ornithology and sonification offered in this chapter have already hinted that the recording and design of sounds are equally important elements of sonic skills. So are the ability to reproduce sounds through physical mimicry (Harris and Van Drie 2015) and to store, retrieve, and circulate sound recordings (Bruyninckx 2013).

These elements, too, can enhance or reduce epistemological authority. To take the case of sonification, many sonification researchers regard the fact that "the traditional carrier of the symbolic knowledge generated by science, paper, hardly begins to meet the requirements of communicating sound" (Dayé and Campo 2006: 360) as a major stumbling block for the scientific acceptance of their techniques. Sound has traditionally been difficult to circulate and integrate with written text, but the development of digital media permitting an easier integration of text and sounds may help sound recordings to catch up, at least partially, with graphical images when it comes to exerting scientific authority (see Supper 2012, 2015). Here, too, tools and sonic skills are closely intertwined.

THE MISSING THIRD DIMENSION: LISTENING TO WHAT?

So far, we have proposed a two-dimensional typology of listening modes, and offered some reflections on how these listening modes—in particular, the ability to shift between different modes of listening—link up with other sonic skills in knowledge-making practices. However, a third dimension of listening has mostly been taken for granted in our analysis: the dimension of *what* it is that scientists, engineers and physicians are listening to. Sidestepping this dimension in our typology was a deliberate decision. Whereas the other two dimensions allowed us to configure a finite number of categories that can nonetheless exhaustively describe the listening practices of scientists, engineers, and physicians, the question of what they are actually listening to opens up an infinite number of possible answers. It defies categorization. That does not, of course, mean that the subject matter to which listeners lend their ear is irrelevant.

On the contrary, it matters a great deal *what* scientists and other practitioners listen to, and it would be a grave mistake to disregard subject matter when discussing listening modes and sonic skills. STS scholar Sophia Roosth has noted that "[s]ound has been used in science to explore and gain direct experience of inaccessible places: to sound the depth of an ocean, the inside of a body, and the furthest reaches of space" (Roosth 2009: 349). We would add that this "direct experience" is often highly mediated by scientific instruments and sonic skills, both of which affect the scientists' very conceptualization of what they are studying. An example from ornithology will help us to show that subject matter and sonic skills are inextricable. The sonic skills involved in listening to, recording, storing, and retrieving sound, we claim, co-define the conception of the objects under study.

In the 1930s, British and American ornithologists struggled immensely with the technical and logistical complexities of making sound recordings of birds in the field. While Cornell University student Albert Brand and his colleagues used a large, sensitive sound camera and testing equipment to capture bird sound, amateur birdwatcher Ludwig Koch and his British companions—ornithologists and recording engineers working for the BBC—used a phonograph recorder and wax disks. What they had in common was the need to move heavily loaded vans around in order to do their recording work. Not only did this affect *where* they could make recordings (the site had to be accessible by road, for instance); it also influenced *what* they recorded. Although the sounds of nature were their primary interest, their microphones also picked up the sounds of modern civilization—as they often discovered, to their dismay, after the fact (Bruyninckx 2013: 64ff).

The two groups came up with their own technical solutions to these challenges, and each solution had its consequences. Choices regarding the type and positioning of microphones, for instance, had not only implications for how they made their recordings, but also for their conceptions of birdsong and for their research findings. The British group carefully installed sets of microphones around the space in which a particular bird was expected to produce its song, and would adjust the sound level of each microphone retrospectively, in the editing process. With a little luck, this resulted in recordings that captured both the bird's song and its environmental sounds, giving the recording an atmospheric touch even though the song was foregrounded. If the bird flew away, however, the whole set-up had to be recreated (Bruyninckx 2013: 71ff).

Whereas the British group worked on BBC nature films that were meant to educate and entertain a wide audience, the ornithologists at Cornell were interested first and foremost in establishing a firm scientific reputation. This may partially explain why they adopted a different approach to recording, making use of a parabolic reflector surrounding a microphone. The surface of the parabola reflected sound waves to a dynamic microphone at its focal point. Focusing the sound waves in this way dramatically increased the input to the recording equipment and concentrated it to at least twenty decibels louder than the sounds not caught by its narrow shape, which amounted to an amplification of about fifteen times (Sellar 1976, cited in Bruyninckx 2013: 73).

The reflector enabled the Cornell ornithologists to pick up bird sound from a considerable distance, making the exact position of the microphones less important and bringing less accessible parts of nature within easier reach. At the same time, microphones with parabolic reflectors staged a "sterile sound," creating a "close-up" of acoustic events. This, argues Bruyninckx, produced not only a form of sound that was half-way between the laboratory and the field, but also "a still-life motif in a clearly demarcated acoustic landscape" (Bruyninckx 2013: 55, 76, 79).

It was only at the beginning of the twenty-first century that ornithologists began to realize their focus on bird sound proper had come at a price. Recent research has indicated that at least one bird species sings at a higher frequency when living in urban areas than the same species living in rural surroundings (Bruyninckx 2013: 151; Slabbekoorn and Peet 2003). For a long time, ornithologists had simply missed this possibility as a consequence of their preference for clean sound. They had treated environmental noise as a disturbance rather than as an informant. Their approach had been both enabled and constrained by their tools, and their carefully crafted sonic skills had affected their knowledge claims.

CONCLUSIONS

This chapter has argued that we need both the notion of listening modes and the notion of sonic skills to understand how sound has been used as a path to knowledge making in science, medicine, and engineering. Its analysis began with a typology of listening practices. Certainly, our project is not the first to present such a typology; typologies of listening abound in scholarly work on sound. Building on such literature and on our own case studies of Western scientists, doctors, mechanics, and engineers, we have distinguished between six modes of listening, operating on two dimensions. Monitory, diagnostic, and exploratory listening refer to different purposes of listening in the sciences; analytic, synthetic, and interactive listening express particular ways of listening. We have also stressed, however, that scientists, engineers, and doctors are not required only to engage in any given one of these modes, but additionally—or especially—to shift between modes. Tools and instruments, whether multichannel stethoscopes, tape recorders, all sonification software, enable particular forms of listening and mode-shifting.

We also pointed out that sonic skills are not limited to listening skills. Although our notion of sonic skills encompasses the skills that experts need in order to employ the various listening modes, it also encompasses the ability to design, record, store, mimic, and retrieve sound. All these sonic skills are associated with the handling of specific instruments—virtuosity in sonic skills means not just the ability to use one's ears, but also the ability to handle various tools and instruments. These practical tools often also play a symbolic role in the knowledge practices of scientists, engineers, and doctors: they can symbolically enhance the epistemological status of listening.

Sonic skills have repercussions on the knowledge claims that can be made in science, engineering, and medicine. The decision to employ a particular technique in the recording of sound, for instance, is not an innocent one: it can affect the substance of knowledge claims that are made and the conceptions of the objects under study. In order to understand the knowledge practices of scientists, engineers, and physicians, then, it pays to consider the listening modes and sonic skills involved in their production. Doing so deepens our insights in the role of sound and listening in the sciences, and might also inspire research into the contribution of other non-visual senses in knowledge making.

As for sound and listening, there is more to discover than we have done so far. We are interested, for instance, in the reopening of debates on the status of sensory information every time new, and epistemologically still unstable, tools are introduced that make the inaudible audible or translate data from one sensory mode into another, as was the case for the spectrograph. We still have much to learn about the conditions under which sonic skills are accepted or contested in knowledge-making practices. These are the topics of the following chapters. But without an understanding of modes of listening, and of their relationship with tools and the skills to handle them, we cannot begin to tackle those issues.

NOTES

1. The sudden increase of interest in listening practices in the late nineteenth and early twentieth century has been linked to anxieties related to the rapidly changing world of music, which had been shaken up by new tuning systems, new tones, and new music theories (Hui 2013), as well as the introduction of new sound technologies such as the gramophone (Maisonneuve 2001).

2. Personal interviews by Alexandra Supper with Christian Dayé (March 17th, 2008) and Alberto de Campo (October 16, 2009).

3. In some cases, it can be difficult to ascertain where the primary literature ends and the secondary literature begins. This is especially true for sonification, as sonification researchers frequently refer to literature by film scholars such as Michel Chion or composers such as Pierre Schaeffer when publishing in sound studies or musicology journals (Vickers 2012; Grond and Hermann 2014).
4. Markendaya, Virat. (2012). Listening to the Sun on a Loop: a Composer Pricks his Ears up for NASA and Helps to Make a Discovery, Scienceline, available at http://scienceline.org/2012/03/listening-to-the-sun-on-a-loop (last accessed August 18, 2017).
5. http://www.vice.com/motherboard/the-space-composer (last accessed on February 20, 2015).
6. Many sonifications are made for monitory purposes, but these usually assume a listener who is too busy with other tasks to actively interact with the sound. If monitory interactive listening is appropriate for car drivers but not usually for sonification users, this is because the sound of a car engine is a by-product, whereas the sound of the sonification is designed as a goal on its own account; a change in the sound of the sonification would usually be explicitly and deliberately caused by the user, whereas a change in the sound of an engine might be a side effect.

REFERENCES

Adorno, T. W. (1977/1962). Typen musikalischen Verhaltens. In *Einleitung in die Musiksoziologie* (pp. 14–34). Frankfurt am Main: Suhrkamp Verlag.
Bijsterveld, K., & Krebs, S. (2013). Listening to the Sounding Objects of the Past: The Case of the Car. In K. Franinović & S. Serafin (Eds.), *Sonic Interaction Design* (pp. 3–38). Cambridge: MIT Press.
Bijsterveld, K., Cleophas, E., Krebs, S., & Mom, G. (2014). *Sound and Safe: A History of Listening Behind the Wheel.* Oxford: Oxford University Press.
Bregman, A. (1994). *Auditory Scene Analysis: The Perceptual Organization of Sound.* Cambridge: MIT Press.
Bruyninckx, J. (2013). *Sound Science: Recording and Listening in the Biology of Bird Song, 1880–1980* (Ph.D. thesis, Maastricht University).
Chion, M. (2005/1990). *Audio-Vision: Sound on Screen.* New York: Columbia University Press.
Dayé, C., & de Campo, A. (2006). Sounds Sequential: Sonification in the Social Sciences. *Interdisciplinary Science Reviews, 31*(4), 349–364.
Dell'Antonio, A. (Ed.). (2004). *Beyond Structural Listening? Postmodern Modes of Hearing.* Berkeley: University of California Press.
Douglas, S. J. (1999). *Listening In: Radio and the American Imagination, from Amos 'n' Andy and Edward R. Murrow to Wolfman Jack and Howard Stern.* New York: Times Books.

Flowers, J. (2005). Thirteen Years of Reflection on Auditory Graphing: Promises, Pitfalls, and Potential New Directions. In *Proceedings of the 11th International Conference on Auditory, Display, Limerick, Ireland, July 6–9* (pp. 405–409).

Gaver, W. W. (1989). The SonicFinder: An Interface That Uses Auditory Icons. *Human-Computer Interaction, 4*(1), 67–94.

Gibling, S. P. (1917). Types of Musical Listening. *Musical Quarterly, 3*(3), 385–389.

Goodman, D. (2010). Distracted Listening: On Not Making Sound Choices in the 1930s. In D. Suisman & S. Strasser (Eds.), *Sound in the Age of Mechanical Reproduction* (pp. 15–46). Philadelphia: University of Pennsylvania Press.

Grond, F., & Hermann, T. (2014). Interactive Sonification for Data Exploration: How Listening Modes and Display Purposes Define Design Guidelines. *Organised Sound, 19*(1), 41–51.

Harris, A. (2015). Sounding Disease: Guest Blog at Sociology of Diagnosis website. Available at https://www.facebook.com/SociologyOfDiagnosis/posts/799049830181091. Last accessed August 18, 2017.

Harris, A., & Van Drie, M. (2015). Sharing Sound: Teaching, Learning and Researching Sonic Skills. *Sound Studies: An Interdisciplinary Journal, 1*(1), 98–117.

Hermann, T. (2002). *Sonification for Exploratory Data Analysis* (Ph.D. thesis, Bielefeld University).

Hermann, T. (2011). Model-Based Sonification. In T. Hermann, A. Hunt, & J. G. Neuhoff (Eds.), *The Sonification Handbook* (pp. 399–427). Berlin: Logos Verlag.

Hermann, T., & Hunt, A. (2005). An Introduction to Interactive Sonification. *IEEE Multimedia, 12*(2), 20–24.

Hermann, T., & Hunt, A. (2011). Interactive Sonification. In T. Hermann, A. Hunt, & J. G. Neuhoff (Eds.), *The Sonification Handbook* (pp. 273–298). Berlin: Logos Verlag.

Hermann, T., Hunt, A., & Neuhoff, J. G. (2011). Introduction. In *The Sonification Handbook* (pp. 1–6). Berlin: Logos Verlag.

Hui, A. (2013). *The Psychophysical Ear: Musical Experiments, Experimental Sounds, 1840–1910.* Cambridge: MIT Press.

Krebs, S. (2012a). Automobilgeräusche als Information: Über das geschulte Ohr des Kfz-Mechanikers. In A. Schoon & A. Volmar (Eds.), *Das geschulte Ohr: Eine Kulturgeschichte der Sonifikation* (pp. 95–110). Bielefeld: Transcript.

Krebs, S. (2012b). "Notschrei eines Automobilisten" oder die Herausbildung des Kfz-Handwerks in Deutschland. *Technikgeschichte, 79*(3), 185–206.

Krebs, S. (2012c). "Sobbing, Whining, Rumbling": Listening to Automobiles as Social Practice. In T. Pinch & K. Bijsterveld (Eds.), *The Oxford Handbook of Sound Studies* (pp. 79–101). Oxford: Oxford University Press.

Krebs, S. & Van Drie, M. (2014). The Art of Stethoscope Use: Diagnostic Listening Practices of Medical Physicians and "Auto Doctors", *ICON: Journal of the International Committee for the History of Technology, 20*(2), 92–114.

Lacey, Kate. (2013). *Listening Publics: The Politics and Experience of Listening in the Media Age*. Cambridge: Polity Press.

Landi, E., Alexander, R. L., Gruesbeck, J. R., Gilbert, J. A., Lepri, S. T., Manchester, W. B., et al. (2012). Carbon Ionization Stages as a Diagnostic of the Solar Wind. *The Astrophysical Journal, 744*(2), 100.

Lorimer, J. (2008). Counting Corncrakes: The Affective Science of the UK Corncrake Census. *Social Studies of Science, 38*(3), 377–405.

Maisonneuve, S. (2001). Between History and Commodity: The Production of a Musical Patrimony Through the Record in the 1920–1930s. *Poetics, 29*(2), 89–108.

Mody, C. C. M. (2005). The Sounds of Science: Listening to Laboratory Practice. *Science, Technology and Human Values, 30*(2), 175–198.

Pinch, T., & Bijsterveld, K. (2012). New Keys to the World of Sound. In *The Oxford Handbook of Sound Studies* (pp. 3–35). Oxford: Oxford University Press.

Rice, T. (2008). "Beautiful Murmurs": Stethoscopic Listening and Acoustic Objectification. *Senses and Society, 3*(3), 293–306.

Rice, T. (2010). "The Hallmark of a Doctor": The Stethoscope and the Making of Medical Identity. *Journal of Material Culture, 15*(3), 287–301.

Roosth, S. (2009). Screaming Yeast: Sonocytology, Cytoplasmic Milieus, and Cellular Subjectivities. *Critical Inquiry, 35*(2), 332–350.

Slabbekoorn, H., & Peet, M. (2003). Ecology: Birds Sing at a Higher Pitch in Urban Noise. *Nature, 424*(6946), 267.

Stockfelt, O. (1997). Adequate Modes of Listening. In D. Schwarz, A. Kassabian, & L. Siegel (Eds.), *Keeping Score: Music, Disciplinarity, Culture* (pp. 129–146). Charlottesville: University Press of Virginia.

Subotnik, R. R. (1991). *Developing Variations: Style and Ideology in Western Music*. Minneapolis: University of Minnesota Press.

Subotnik, R. R. (1996). Toward a Deconstruction of Structural Listening: A Critique of Schoenberg, Adorno, and Stravinsky. *Deconstructive Variations: Music and Reason in Western Society* (pp. 148–176). Minneapolis: University of Minnesota Press.

Summers, E. (1916). Notation of Bird Songs and Notes. *The Auk, 33*(1), 78–80.

Supper, A. (2012). *Lobbying for the Ear: The Public Fascination with and Academic Legitimacy of the Sonification of Scientific Data* (Ph.D. thesis, Maastricht University).

Supper, A. (2015). Sound Information: Sonification in the Age of Complex Data and Digital Audio. *Information & Culture: A Journal of History, 50*(4), 441–464.

Supper, A., & Bijsterveld, K. (2015). Sounds Convincing: Modes of Listening and Sonic Skills in Knowledge Making. *Interdisciplinary Science Reviews,* 40(2), 124–144.

Thorpe, W. H. (1958). The Learning of Song Patterns by Birds, with Especial Reference to the Songs of the Chaffinch 'Fringilla Coelebs'. *The Ibis, 100*(4), 535–570.

Truax, B. (2001/1984). *Acoustic Communication* (2nd ed.). Westport: Greenwood.

Van Drie, M. (2013). Training the Auscultative Ear: Medical Textbooks and Teaching Tapes (1950–2010). *The Senses and Society, 8*(2), 165–191.

Vickers, P. (2011). Sonification for Process Monitoring. In T. Hermann, A. Hunt, & J. G. Neuhoff (Eds.), *The Sonification Handbook* (pp. 455–491). Berlin: Logos Verlag.

Vickers, P. (2012). Ways of Listening and Modes of Being: Electroacoustic Auditory Display. *Journal of Sonic Studies, 2*(1). Available at http://journal. sonicstudies.org/vol02/nr01/a04. Last accessed August 18, 2017.

Walker, B. N., & Nees, M. A. (2011). Theory of Sonification. In T. Hermann, A. Hunt, & J. G. Neuhoff (Eds.), *The Sonification Handbook* (pp. 9–39). Berlin: Logos Verlag.

Williams, S. M. (1994). Perceptual Principles in Sound Grouping. In G. Kramer (Ed.), *Auditory Display: Sonification, Audification, and Auditory Interfaces* (pp. 95–125). Reading: Addison-Wesley Publishing Company.

Worrall, D. (2009). *Sonification and Information: Concepts, Instruments and Techniques* (Ph.D. thesis, University of Canberra).

Zwikker, C. (1934). De oorzaken van het geluid bij automobielen. In Anonymous, *Verslag van het 'Anti-lawaai Congres,' georganiseerd te Delft, op 8 november 1934 door de Koninklijke Nederlandsche Automobile Club in samenwerking met de Geluidstichting* (pp. 70–77). Delft: KNAC/Geluidstichting.

CHAPTER 4

Resounding Contestation:
The Ambiguous Status of Sonic Skills

Abstract This chapter asks how listening in the sciences became contested over time. Why did sonic skills, and notably diagnostic analytic listening, acquire such an ambiguous epistemological status? The chapter traces the rise of mechanical and visual technologies such as the spectrograph, and the shifting relationships of trust between makers and users of knowledge. It shows how each novel knowledge-making technology, either auditory or visual, requires processes of sensory calibration with existing technologies. And it discusses how sonification scientists have strategically presented visualization as both ally and enemy for trained ears, without yet finding a "killer application".

Keywords Epistemological contestation · Sensory calibration · Trust between knowledge makers and knowledge users · Trained ears

TWITTERING TIMBRADOS

On January 23, 2015, a Maastricht University lecture hall featured four Timbrado canaries, two Edison phonographs, a vintage gramophone, a serinette, a piccolo player, an artist-researcher, and an audience in eager anticipation. The artist-researcher, Aleks Kolkowski, intended to reenact and demonstrate how bird sound was recorded by bird researchers and the phonograph industry in the early years of the twentieth century. It was one

© The Author(s) 2019
K. Bijsterveld, *Sonic Skills*,
https://doi.org/10.1057/978-1-137-59829-5_4

87

of the events we organized in the context of our Sonic Science Festival, an outreach activity accompanying the Sonic Skills research project.

Much went "wrong" during the demonstration. It was not that the canaries did not make themselves heard. We had expected them to remain silent, as canaries normally do not sing in January. With this in mind, we had brought along a serinette, a mini-organ originally employed by eighteenth- and nineteenth-century bourgeois bird lovers to seduce or instruct birds to sing.[1] But our Timbrados were trained birds and were used to twittering under the most stressful conditions, such as the contests their owners sent them to. They responded to the slightest high-pitched sounds, and burst out in loud concert when one of us played the serinette.

Recording their sounds with a mechanical, early twentieth-century phonograph—without the use of microphones—turned out to be a much harder nut to crack. In addition to the phonographs we had available, Kolkowski had brought a portable oven to soften the wax on a cylinder, two different needles (one for recording and one for replaying), and a horn to capture the sound produced by the birds. That sound's acoustic energy had to make a membrane vibrate and move the needle, which would leave its traces in the wax. Kolkowski shouted the date and place of the event into the horn, as early twentieth recordists would have done. Unfortunately, when the cylinder was replayed, his voice sounded too high and his words too fast—he had cranked up the phonograph too hard, making the cylinder run too fast. By explicating this, Kolkowski demonstrated the had listened diagnostically in order to understand the disappointing quality of the recording.

Even worse, the cylinder replayed the bird sounds only faintly, if not at the very threshold of hearing. Apparently, the sound waves had not reached the membrane with sufficient energy. Nor did Kolkowski know for certain whether the wax had been heated to the right temperature. So whereas the audience heard the birds sing quite loudly, the phonograph had "heard" hardly anything. As Kolkowski explained, this happened often in the past as well, which is why early recordists placed caged birds inside the horn in order to capture their sounds. Because the horn plunged the birds into darkness, however, they often refused to sing. This inspired the recordists to bring bird impersonators or flautists to the recording studios, just in case. We had a piccolo player, Anne Davids, as fallback option, and she beautifully played transcriptions of bird sounds as well as the flute score from Olivier Messiaen's "Le Merle noir"

(The Blackbird), originally composed for flute and piano. This time, the phonograph succeeded in recording and audibly replaying the music we had listened to. We recognized the tunes.

Although the demonstration had partially failed, the event was a success in terms of reenacting early twentieth-century bird recording. It conveyed to the audience what had been at stake at the time and articulated the sonic skills that were involved: heating the wax to the right temperature, turning the cylinder in such a way that it would run at the right speed (leading to the right recording pitch), adjusting the distance between sound source and phonograph, using the correct needles, and having the recorded subjects behave in preferred ways. Listening to birds through phonographic recording clearly entailed an intriguingly wide range of embodied forms of knowledge.

Showing this complexity of sonic skills in a performance for recordist and sounding subjects, the demonstration also underpinned Jonathan Sterne's claim that recording "is a form of exteriority: it does not preserve a preexisting sonic event as it happens so much as it creates and organizes events for the possibility of preservation and repetition" (Sterne 2003: 332; see also Bronfman 2016: 228; Brady 1999: 6–7). Sonic examinations, like other ways of staging phenomena to be recorded, redefined the researching experts in terms of the skills required, but they also redefined the objects studied. Things, animals, and humans were made to sound loud enough to be captured by the vibrating needles, membranes, and amplifying tools of their time. In some cases, subjects and researchers closely co-operated in creating the recordings. As Erika Brady has illustrated for ethnology, the humans under study at times only collaborated with researchers in ways that allowed them to remain true to their own cultural and epistemic conventions, thus impacting on what was recorded (Brady 1999: 111–117). The weight of the recording instruments and means of transportation affected which sounding objects could be reached. Different microphones resulted in different permeations between sounding objects and their environment, with huge effects on the questions posed, as we saw in the previous chapter. The sonic traces of those objects, and the resulting issues of interest were, as I will demonstrate below, even co-defined by the maximum length of the sound recordings or their visualizations.

In ornithology, mechanical sound recording did not remain the preferred medium for capturing and analyzing bird vocalization for long. Soon, ornithologists would rank the epistemological value of mechanically visualizing sound more highly than that of audio recording plus manual notation, at least for "diagnostic" purposes as set out in the taxonomy of listening presented in Chapter 3. In automotive engineering, mechanical visualization replaced embodied listening as the portal to systematic diagnosis, earlier in the United States than in Germany. And in medicine, the situation differed for different specializations. How should this shifting balance between the senses be interpreted? Switching between sensory modalities, according to STS scholars Regula Burri et al. (2011), may be considered "an epistemic tactic with which the different senses are put into productive relations" (p. 4), but this does not yet account for the differences in when and how experts in particular fields switch from one sense to another.

Varying relations of trust between knowledge makers and knowledge takers will be one of the factors; the ecology of diagnostic instruments in which new technologies intervene is another. By following cases over time, I am able to show how, and under which conditions, practices of listening were critiqued or replaced by other techniques of knowing in science, medicine, and engineering. In terms of the modes of listening discussed in the previous chapter, the most explicit and heated discussions focused on diagnostic listening as opposed to monitory or exploratory listening. As I will explain in Chapter 5, this does not imply that the aims of monitory and exploratory listening diminished in significance in the sciences over time—quite the contrary. But attempts to mechanize the analysis of data through visualization centered on diagnosis.

Replacing embodied listening to sound by automatically recording sound, and subsequently by the mechanical visualization of sound, I will additionally claim, required processes of sensory calibration.[2] That is, new sensory practices had to be "anchored" in existing ones in order to acquire authority. Just as some music and dance researchers had tried to enhance the reliability of the manual notation of sound by calibrating it with the embodied, kinetic experience of music making and dancing, so the promoters of automatic registration and visualization of sound aimed to augment the authority of these practices by materially and metaphorically calibrating it with known sensory knowledge practices.

Such attempts at calibration were not always successful. The sonification community, for instance, hovered between presenting data visualization as a practice with which sonification had much in common and as one from which it should depart. But this rhetorical flexibility did, as yet, nothing to help the community find a convincing "killer application".

Beyond the Infallible Ear:
Repeated Listening and Sensory Calibration

In Chapter 2, I discussed several scholarly fields that (with more or less initial hesitation) embraced the phonograph, most notably for its mimetic and mnemonic functions. In 1931, the Romanian folklorist Constantin Brăiloiu recapitulated these advantages when he expressed his trust in the phonograph's recording diaphragm as "infallible ear" (p. 394) in the essay "Outline of a Method of Musical Folklore":

> The concern for objectivity impels us, first of all, to undertake the mechanical recording of melodies. Only the machine is objective beyond question and only its reproduction is unquestionable and complete. No matter how well we notate a performer's melody by dictation, we will always miss something in our notation, whether it be the timbre of the voice of that particular coloration of the melody due to the vocal production of the peasant, [or] ... the timbre of the instruments. Furthermore, the mechanical recording avoids fatiguing the informants, and facilitates an extensive collection. Finally, it provides us with a means of control which no exact science can do without. (Brăiloiu 1970/1931: 393)

By asserting the accuracy and reliability of mechanical sound recording, Brăiloiu voiced the same ideals as the ornithologists had done. Similarly, creating extensive collections and boosting productivity through phonographic recording was no less important in ornithology as it was in folklore research and ethnology (Brady 1999: 67). The only difference was that the naturalists did not have to worry about fatiguing their informants. Birds in the field could not care less.

Mechanical sound recording was not free of problems, however. I touched on some of these when discussing ornithology: the need for heavy equipment, motorized transportation, and electricity affected the recordability of the field, and the distance between bird and microphone in natural settings generated a dichotomy between atmospheric and close-up recordings. The folklorist Brăiloiu also mentioned that wax cylinders crumbled easily and deteriorated quickly; they did not "withstand more than twenty to twenty-five playings" and thus had to be transferred to discs if possible. He also thought it wise to have the infallible ear collaborate with "the infallible eye of the lens." The engravings in wax or ebonite might "keep a flicker of life" and be "easier to comprehend" if auxiliary documentation such as photographs, films, and index cards with detailed descriptions of the singers and their ritual events were stored as well (Brăiloiu 1970/1931: 393–394, 397). For Brăiloiu, rural music was especially interesting as a living tradition, "the dual effort of integration and adaptation which tends, on the one hand, to pour the attributes of modern civilization into the mold of the tradition, and on the other hand, to impose upon this tradition the appearance of the contemporary world" (p. 392).

At least in Brăiloiu's view, producing knowledge from sound recordings meant aligning and anchoring them in visual documentation beyond musical notation. Such forms of material sensory calibration were also crucial in ornithology. Joeri Bruyninckx has shown how the Cornell Library of Natural Sounds attempted to make volunteers' recordings as valuable to research as professionals' recordings. This aim acquired particular significance in the 1950s, when the magnetic tape recorder—less heavy, expensive, and cumbersome to use than the gramophone disc-cutter—became available to amateur sound hunters, while ornithologists' growing interest in population ecology and ethology made them keen to collect the greatest possible variety of bird recordings. The Library tried to ensure the accuracy and reliability of recordings by asking amateurs, for instance in its *Bioacoustics Bulletin*, to fill out standardized forms with information on the recorded bird species as well as the date, place, and other conditions of recording, and to use pitch pipes with a standardized tone of 440 Hz. Taping this tone before the bird's vocalization "enabled future users to calibrate their playback equipment and to detect deviations in recording speed" (Bruyninckx 2015: 358). Ornithologists thus tried to materially calibrate sound recordings both through existing sounding instruments, such as pitch pipes, and through the existing visual format of the standardized form. Analytic and synthetic listening for comparative diagnostic purposes was only considered possible through calibration and standardization of recording practices.

Encouraging amateurs to contribute to sensory calibration was only one aspect of the exchange relation that the Library developed with its volunteers. The institution shored up that relationship with various forms of capital. Social capital was exchanged when advice on making field records and expertise on the recorded birds was offered in return for the moral obligation to share information with the Library. Sponsoring field trips with travel funds and equipment in return for the recordings and the associated copyrights was an exchange of economic capital. And granting the amateurs authorship of the recordings and the distinguished status of research associates brought symbolic capital into play. Bruyninckx (2015) has explained in these Bourdieusian terms how the Library secured volunteers' prolonged investment in its undertaking.

The sheer number of sound recordings collected, and the time it took to listen to them repeatedly, also prompted scientists to ask what exactly the recordings added to analysis beyond mere documentation. In this, they showed similarities with another group of professionals. As media scholar Tom Willaert has shown, several literary writers and intellectuals in late nineteenth- and early twentieth-century Flanders and the Netherlands felt the need to explicitly define their creative contribution now that the phonograph had proved capable of capturing verbatim language—what was left for writers to do? One such writer, the Dutch novelist Multatuli (that is, Eduard Douwes Dekker), welcomed the phonograph as the perfect embodiment of his own poetics: his aspiration to natural-sounding language and an associatively unfolding argument. Others saw phonographic writing as a form of mechanical registration that might be allowable for entertainment literature, but must be kept out of high literature at all costs. Instead, writers should interpret and critically reflect upon what they observed; use exemplary language; and elaborate, contrast, and synthesize at proper places in their texts (Willaert 2016: 1–22, 33–41).

Scientists and scholars, too, were eager to underline what their expertise could add to mere registration. Rather than highlighting their subjectivity, as some literary writers had done, most cited their capacity to compare sounds systematically through repeated listening. Playing recordings at reduced speed was considered helpful in this process. In the 1960s and 1970s, several bird sound recordists tried to support the epistemic authority of this technique, in their case executed with gramophones, by comparing recorded sounds with microscopic images— just as the ethnomusicologist Benjamin Ives Gilman had done decades earlier when writing about "magnifying" the sound of his phonograph recordings. The technique allowed ornithological recordists to extend sonic fragments, so they claimed, just as microscopy enabled scientists to enlarge visual details (Bruyninckx 2018). Such a rhetoric intervention entailed a metaphorical rather than a material calibration with a known visual instrument and its sensory enhancement. Whether material or metaphorical, however, the processes of sensory calibration had to improve the reliability and authority of sound recording in the fields that used it.

AT A GLANCE: MECHANICAL
VISUALIZATION AND PROFESSIONAL AUDITION

The world first learned about the sound spectrograph in 1947. That was the year when Ralf Potter, Director of Transmission Research at Bell Telephone Laboratories, and his former colleagues George Kopp and Harriet Green published their report *Visible Speech*. Although the title expressed their primary interest in visualizing speech, the spectrograph permitted the visualization of sound more generally. Technologies for the transduction of sounds into images as such were by no means new. The nineteenth-century "oscillograph," for instance, was already able to visualize the frequency and intensity of pure tones across time, with intensity represented as the amplitude of the sound's horizontally developing waveform.[3] But the spectrograph enabled its users to make images of complex tones or a spectrum of frequencies—the default situation in speech. These images, or "sound spectrograms," plotted time horizontally and frequency vertically, and displayed the intensity of sound as shades of darkness.[4] The assumption was that the sound spectrograph mimicked human sound perception more accurately than earlier instruments had done because the ear also processes sound in terms of different frequency components (Potter et al. 1947: 8–13; Mills 2010: 38). Potter and his colleagues seemed to reserve the word "sonogram" for the speech spectrogram, but other early users distinguished less clearly between the two terms. Today, "sonogram" predominantly refers to ultrasound images in medical practices.

The *Visible Speech* authors believed the sound spectrograph would be useful in teaching the deaf and hearing-impaired to speak by giving them feedback on the quality of their speech, and would enable them to understand telephone calls by reading sonograms of telephone conversations in real time, though they admitted to having no conclusive proof of whether this was really possible (p. 6). They also referred to military usages of the technology. Behind the scenes, the military was working on uses of the spectrograph for unscrambling telephone messages and identifying speakers through "voice printing" (Fehr 2000; Broeders 2002; Mills 2010: 52). At the end of *Visible Speech*, the authors speculated about an impressive series of possible future uses, including options for research in bioacoustics:

A little experience with the patterns of animal, bird, and insect sounds suggests that there may be fascinating possibilities of analysis, illustration and
discussion that have not been available in the past. For example, perma-
nent patterns of bird songs can be examined in great detail, and there is
reason to believe that song differences within one group could be readily
identified by visual comparisons. In fact, it would not be surprising if the
song habits of individual birds could be recognized by close examination
of the patterns. ... The song becomes a signature!

If detailed analysis of song patterns is possible, there would seem to be
a wide new field of study open to the ornithologist. Perhaps bird books
and periodicals of the future will be filled with song pictures, and serious
readers may become well enough acquainted with this sound language to
read visible patterns of bird music in the way a musician reads a musical
score. (Potter et al. 1947: 410–411)

From the late 1950s, ornithological journals would indeed be full of
sound spectrograms. In the early years of that decade, the ethologists
William Thorpe and Peter Marler had started experimenting with the
technique, which was quickly taken up by the world of ornithology. In
contrast to mechanical sound recording, the spectrograph could only
represent a few—two to four—seconds of sound. Nevertheless, it was
welcomed for its capacity to visualize pitches far beyond human hearing and to cope with the high speed and frequent modulations of bird
vocalizations (Bruyninckx 2013: 119ff). In fact, it was these characteristics that fostered ornithologists' interest in studying bird calls—short
in length, high in pitch, potentially rapid in repetition—instead of more
extended bird vocalizations, another example of the knowledge effects of
new instruments. The spectrograph also contributed to a focus on variation among other short elements in the repertoires of birds: did such
variation signify new species, learned behavior, functional characteristics,
adaptations to environmental change, or crucial forms of communication
(Bruyninckx 2013: 160–161)? Looking back in 2004, Peter Marler—
probably unintentionally—copied the rhetorical strategies of some of his
predecessors in ornithology by comparing the sound spectrograph to the
microscope (Marler 2004, cited by Bruyninckx 2013: 119). His comparison was yet another instance of metaphorical sensory calibration, as well
as an expression of a further version of mechanical objectivity—leaving
the ear out of a job not only in the field, but also in the office where the
transcription phase took place.

Meanwhile, ethnomusicologists preferred the melograph to the spectrograph, because it produced two line graphs, one representing pitch across time and the other volume across time, creating an easily readable visualization of melodic, and especially vocal melodic lines. The melograph's inventor, the American musicologist Charles Seeger, had imagined it in the 1930s, but realized its potential only in the 1940s thanks to Potter's technical drawings of the sound spectrograph, and had its first model built in the 1950s (Mundy 2018: 133; Prescatello 1992: 212). Despite the differences between the melograph and the spectrograph, both instruments produced visualizations of sound that facilitated the distribution of research data through publications on paper, as well as fruitful combinations of spectrograms with other forms of visual representation. Seeger advised his peers to add musical notation to the graphs, for instance, so that the notation would *prescribe* to readers how to sing a particular song while the graphs would *describe* how it had actually been performed (Prescatello 1992: 212–217; Mundy 2018: 215). And as Bruyninckx explains with the help of Bruno Latour's (1986) work on inscriptions, such images were mobile, immutable (or at least less vulnerable to erasure and alteration than sound recordings), flat, and thus easily "overseen, cut up, scaled, recombined or superimposed" (Bruyninckx 2013: 144). Whereas listening to sound recordings required a substantial time investment for each playing, sound spectrograms could be transformed into optically consistent images that could then be synoptically presented so as to compare the images at a glance and cascade them into other abstractions, such as numbers.

Bruyninckx adds, however, that the sound spectrograph did not achieve "instantaneous intelligibility" (2013: 121). Some ornithologists began to abstract the sound spectrograms into calligraphic signs that they thought were more easily readable and printable and better represented the essence of the patterns observed. Masking the spectrograph's visualizations of background noises with white paint was part of that procedure. Others expressed their dislike of such interventions—or indeed any interventions. Still others added verbal descriptions and syllabic notations to the sound spectrograms, hoping to give each other "an impression of how the sounds might appear 'to the human ear.'" Such subjective accounts, Bruyninckx claims, did not only supply "information that could not be conveyed otherwise"; they also marked "a perceptive minimum to orient the observations of other ornithologists" (2013:

135). Offering such manually notated aural information thus assisted the analysis of bird sound by peers, and again anchored a new visual technology in earlier techniques.

Issues of legibility and analysis were also raised by ornithologist Donald J. Borror in 1956. He noted that it was very hard to define the beginning and the end of Carolina wren song phrases just by looking at sound spectrogram. Because delimiting phrases was crucial for his work, he had to combine the imagery with listening to sound recordings at reduced speed. In the early 1970s, Robert Lemon, a biologist working at McGill, relied on the auditory sense in the same way. In his study of how cardinals responded to the prerecorded songs of other cardinals, he mentioned that although all new recordings of their vocal behavior had been analyzed with help of a sound spectrograph, "much information ..., especially relating to the sequences of different song types, was gathered by listening to the birds sing and then recording the data in a notebook" (Lemon and Chatfield 1971: 1, cited by Bruyninckx 2013: 135). In this way, both Borror and Lemon underpinned the understanding of sound spectrograms with analytic listening.

Some ornithologists even started combining sound spectrograms with musical notation. A few did so because they refused to abandon the search for musical patterns in bird vocalizations, a search that was by then highly contested. Others argued that the linear frequency scale of sound spectrograms did not express the logarithmic sensation of pitch to which both humans and birds responded, or else considered musical scores more accessible and comprehensible than spectrographic images.

For these ornithologists, musical skills remained an integral part of the ornithologist's expertise, or, following Bruyninckx's (2013: 41–42) reference to Thomas Porcello, of the ornithologist's "professional audition." Thomas Porcello coined this term in 2004, based on Charles Goodwin's 1994 notion of "professional vision," to capture the discursive and embodied competences of experienced sound studio engineers—expressed in their shared technical, musical, and linguistic repertoire for bringing together audio technologies, techniques, and sonic ideals—and to emphasize novices' lack of such authoritative and efficient expertise.

At another level, Bruyninckx argues (2013: 134), abstracting spectrograms, adding manual notations to automatically generated spectrographic images, and carrying out listening exercises meant the introduction to ornithology of what Lorraine Daston and Peter Galison

(1992, 2007: 309–361) have called "trained judgment." Trained judgment represents the gradual enrichment of the ideal of mechanical objectivity in the sciences by confidence in the sophistication of experts' interpretive skills, drawing on an intuitive and holistic understanding of data patterns developed through years of experience. But it also entails judgment of when exactly such interpretive skills should inform analysis and when they should not.

Reintroducing musical notation, for instance, was considered unacceptable by the large majority of ornithologists. True, some of them conceded, the composer Messiaen had successfully simulated bird song with such notations—but using them in biology would mean wrongly assuming that birds were musical creatures (Bruyninckx 2013: 141).

METER READING AS A TECHNOLOGY OF TRUST IN EXPERT–CUSTOMER RELATIONS

Although sound spectrograms made their presence felt in bird books and periodicals in the 1950s and after, by then ornithology had already experienced a golden age of sound recording—including in commercial terms, with large sales for the records issued by the Cornell Library of Natural Sounds.[5] Other branches of bioacoustics, as well as ethnomusicology, ethnology, and linguistic fields such as dialectology and phonology, also embraced sound recording as a token of advanced scholarship for a substantial period of time.

In the "stethoscopic" fields of medicine and automotive engineering, mechanical sound recording acquired different roles. In medicine, gramophone records enabled physicians to document and circulate rare cases, and were also used in teaching contexts. As early as 1930, a renowned American textbook on physical diagnosis referred to a series of Columbia records that students could listen to repeatedly. But the medical field did not unanimously embrace listening to the body through loudspeakers. In the 1940s, several authors asserted that the technology had a detrimental effect on the isolating and filtering experience of listening with the stethoscope, as background noise and loudspeaker buzzing interfered, destroying—in Van Drie's rendering of these comments—"the impression of the closed acoustic pathway" (2013: 178) that had been the stethoscope's great advantage. One solution was to have a professor in a lecture hall carry out a auscultation live with an electronic stethoscope, then broadcast it to students who had individual stethoscopes plugged into the transmission system, thus reinstating the closed pathway experience. Another was to publish "teaching tapes" (or audiocassette recordings in the 1970s) specially designed to be listened to by holding the stetho-scope's bell a few inches from the speaker of the tape recorder. Employed in this way, the recorder acted as an "electronic chest" (p. 180).

Even so, mechanical sound recording was less widely used for diag-nostic listening in medicine than in fields such as ornithology, ethno-musicology, and phonology. It is no surprise that medicine differs from areas in which sound, as music or language, is the key object of study. But for ornithology, the dissimilarity with medicine is less self-explan-atory. To understand it, we need to take the field-specific ecology of diagnostic instruments into account. Most medical specializations could draw on both the stethoscope and visual diagnostic techniques such as radiology before mechanical sound recording entered the scene. This made mechanical sound recording less essential for diagnostic listening, although it was not entirely absent. In the mid-1960s, for instance, slow-speed magnetic tape recording was used for the objective assessment of "cough suppressants under clinical conditions" (Woolf and Rosenberg 1964); three decades later, visualizing high-speed magnetic tape record-ings of cough sounds was ascribed "considerable value in identifying mechanisms of airway pathology present in respiratory diseases" (Korpáš et al. 1996: 261). Indeed, biomedical acoustics has become a highly developed field. For the most part, however, it focuses on visualizing ultrasound rather than sound in the human auditory range. So while the stethoscope lived on in the ward, and mechanical sound recordings played a modest role in medical educational settings, it was imaging that dominated diagnostic work in behind-the-scenes hospital laboratories.

Stefan Krebs has shown how the transition from auditory to visual diagnostics unfolded in car mechanics in Germany and the United States. As discussed in previous chapters, German mechanics modeled their diagnostic listening skills on the medical world, using automotive stethoscopes as supplements to screwdrivers and listening rods to focus their listening. When confronted with a repair crisis in the 1930s, they did not follow their US colleagues in introducing meters and gauges to restore a relation of trust with customers, but embedded car mechanics in a system of certified guilds. Whereas in the United States, the dam-age to customers' belief in the capacities of mechanics had been miti-gated with legible, and thus visual, instruments as technologies of trust, in Germany the long-standing tradition of crafts guilds underpinned mechanics' automotive authority. Diagnostic listening could become an insignia of the German mechanics' profession, and came to stand "metonymically for all embodied sensory skills car mechanics developed through training and expertise" (Krebs 2014: 355; see also Krebs and Van Drie 2014: 100). After World War II, the American military seized the opportunity to replace the restrictive German guilds by freedom of trade in the zone they controlled. Indicating the deep roots of the guilds in German society, however, the West German government reestablished the system in 1953.

Because of this history, new diagnostic instruments entered the German garage some twenty years later than the American one. Germany's leading manufacturer of car systems, Robert Bosch, had introduced a test instrument for spark plugs as early as World War I, but only used this and similar devices for its own services. When it did start selling diagnostic tools to German repair shops in the 1950s, the dominant argument was not that the devices would increase demand for repairs, as had been the US selling strategy, but that they would save time by helping mechanics to find the cause of troubles without having to disassemble the car. This efficiency argument was important because the German repair world was struggling with a shortage of car mechanics, who could earn much more in the rapidly expanding automotive industry than in repair shops. Other arguments focused on the automotive technology itself—electrical systems being increasingly sophisticated and more highly powered machines permitting smaller tolerances—and on a new form of after-sales service that was geared towards preventing car problems rather than solving them. One article in an automotive journal additionally played the objectivity card. It deployed the medical metaphor again, but this time to advocate visual instead of audible tools: American diagnostic devices were superior to the mechanic's senses, and the tools were "like medical instruments in an operating room, covered in white enamel and chrome, to be wheeled silently towards the patient: the 'sick' automobile" ("Sie fragen," 1957: 43–44, cited by Krebs 2014: 367).

From the late 1960s onwards, car manufacturers offered their garages diagnostic test stands and increasingly required them to be used. In parallel, the trade press criticized traditional methods: "You can no longer master modern automobiles with your expert senses; only with modern diagnostic instruments is it possible to do the necessary tests and adjustments" ("Prüfen," 1966: 20, cited in Krebs 2014: 370). Bosch also pitted conservative experts against progressive ones, who preferred "measuring instead of guessing, checking instead of trying, and testing instead of sensing."[6] Such rhetorical moves towards objectivity, Krebs argues, were entangled with a wider visual culture. Bosch presented the oscilloscope (an oscillograph with a screen), for example, as the mechanic's television (Krebs 2014: 372).

German mechanics did not give in easily, however. They resisted by claiming that American-style diagnostics would produce mechanics who were able merely to change parts, not to truly repair a car. To them, sensory skills were more than filing, drilling, lathing, forging, and listening: they also encompassed tidiness, punctuality, meticulousness, and care. Such skills were considered crucial in mechanics' relationship with customers. An expert mechanic would, for instance, be able to step beyond the standards imposed by manufacturers and "increase the specified valve clearance by another 0.10 mm to avoid valve ticking that often annoyed drivers" (Krebs 2014: 376). Moreover, they had complaints, at times justified, about the accuracy of the equipment, or simply wanted to prevent customers from noticing their inexperience in handling the devices. Only in the late 1970s did the situation begin to change, due to the rise of the electronically rather than mechanically functioning car, which strengthened the position of the manufacturers and repair shop employers at the expense of the mechanics' jurisdiction over embodied skills.

Understanding such context-specific appropriation of visual diagnostic techniques can contribute to the present-day STS debate on tacit knowledge. Sociologist of science Harry Collins distinguishes between three forms of tacit knowledge: relational, somatic, and collective. Relational tacit knowledge refers to knowledge that is not explicit, formalized, or captured in rules, though only for the time being—once social relations change, such tacit knowledge may become explicit after all. Secrets are a case in point. Somatic tacit knowledge is embodied knowledge, hard to explain because of its incarnated character, but possibly explicable in the future; an example is riding a bike. Collective tacit knowledge, the most robust form of tacit knowledge, is knowledge that entirely resists explication because it can only be acquired by spending time with those who already have it; cycling in local traffic, for instance (Collins 2013). Rather than classifying tacit knowledge substantively as Collins does, Michael Lynch finds it more interesting to "examine what is *done* with the notion of 'tacit knowledge'" as a polemical and professional resource (Lynch 2013: 58).

Indeed, this and the previous chapters have shown how German car mechanics succeeded for quite some time in defining their auditory and other sensory expertise as knowledge that could only be learned by spending time with a senior member of their field, so as collective tacit knowledge. The auto-stethoscope was a token of their professional autonomy until labor shortages, diminishing tolerances, ideals of preventive care, and electronically steered cars arrived and there was less and less to be heard.[7]

THERE IS MORE THAN MEETS THE EYE: STRUGGLING WITH TRAINED EARS IN SONIFICATION

The community of researchers who have been promoting data sonification since the early 1990s have a deeply ambiguous relationship with the phenomenon of data visualization. As Alexandra Supper has shown, sonification researchers have passionately lobbied for the ear by positioning sonification as a much-needed *alternative* to visualization. Auditory displays of data, so they commonly claim, allow for an easier recognition of patterns than visual data presentations. But they have also stressed sonification's *similarity* to visualization: just as graphs and diagrams are conventional representations of data that have gained authority over time, sonification deserves, as a convention-in-the-making, to gradually acquire acclaim in the academic world. For sonification enthusiasts, visualization is thus both a phenomenon to depart from and to set as example (Supper 2012a: 264; 2016).

An often used argument for sonification is that the huge amount and widespread availability of digitalized data in science and society today call for new ways of processing. So far, "data exploration tools" have been predominantly "visual in nature, including graphing and plotting software, modeling programs, and 2[D] and 3D visualization software," notes psychologist and computer scientist Bruce Walker, but these tools "fail to exploit the excellent pattern recognition capabilities of the human auditory system, and they also continue to exclude students and researchers with visual disabilities" (Walker 2000: 16f, cited in Supper 2012b: 17). In the words of system biologist Peter Larsen, who has translated data from microbial ecology into a jazz composition, "there is only so much" that a person or even a computer "can do to see patterns in these outrageously huge data sets" (Larsen cited by Brannen 2013: 1, in turn cited by Supper 2015a: 441).

Makers of sonifications consider auditory data display to be especially relevant for those sciences that already work with data based on vibration or oscillation, such as seismology, volcanology, or astronomy because these data are reasonably easy to transfer to the human auditory range. Sonification is also presented as very interesting for sciences that produce data with a temporal dimension, for example electroencephalograms (EEGs) of epileptic seizures. Sonifications of EEGs, their proponents argue, have rhythmic and spatial dimensions that may make the dynamics of seizures more intelligible (Supper 2012b: 14–15).

Given the presumed analytical advantages of listening to data, members of the sonification community find it unfair that the ear has been taken less seriously than the eye in the production of knowledge. This makes them keen to refer to auditory activities they see as precursors to their work. In doing so, they make no explicit distinction between listening to sounds that occur "naturally or as an unintended byproduct of another activity" and to sounds that have been "deliberately made for the purposes of revealing information" (Supper 2015a: 445). Evidently, listening to the second type of sound is closer to sonification proper, but it is a rhetorically effective move to bracket both types of sound together. The use of stethoscopes in medicine and automotive engineering is a long-standing example of the first kind of listening; an instance of the second kind is the Geiger counter. Both cases are often referred to in sonification literature. Another example enthusiastically embraced by the sonification community is psychoacoustician Sheridan Speeth's 1960s digital transposition of seismic signals into sound "in an effort to find a reliable method to distinguish earthquakes from underground nuclear explosions" (Supper 2015a: 446, see also Volmar 2013). In all these cases, historical examples of listening are cited to undermine the presumed dominance of visual modes of analyzing and presenting data.

Proponents of sonification do not always write antagonistically about visualization, however. In their arguments, visualization is sometimes also an authoritative practice prefiguring and acting as a model for sonification. Thus, the claim that "subjective decisions are widely accepted in data visualization" may be used to argue that such interventions are also acceptable in sonification (Supper 2012c: 32). Whereas some members of the sonification community, notably psychologists, ask for quantitative user tests to substantiate "what the average listener actually hears in a sonification," more theoretically inclined sonification researchers deplore such demands (Supper 2012c: 31). Visualization, they point out, is commonly used without scientists requesting user tests showing that it actually works in conveying information, so why should user tests be necessary for sonification? If visualization has organically developed into an accepted set of conventions, why not grant the same evolution to sonification? As one of these promoters has it, the "first visualizations of molecules were not evaluated, they were just made. And they were extremely functional" (Interview Florian Grond, cited in Supper 2012a: 263).

Their ideal is to rely on what Supper calls the "trained ears" of sonification designers and the experts, or domain scientists, whose data are sonified. Supper's formulation plays with Daston and Galison's notion of "trained judgment." In the sonification community, it stands for the position that mapping data onto particular sound parameters always implies choice and subjectivity, but that this is no problem as long as sonification designers and domain scientists are willing to listen to different mappings of the same data sets, which represent different "(sonic) views," to quote Thomas Hermann, a prominent member of the community, and his co-authors (Hermann et al. 2007: 467 cited in Supper 2012c: 32).

Views on visualization among sonification researchers experts have also changed over time. Whereas several of the sonification movement's pioneers expressed the hope that sonification would replace visualization in the long term, most sonification researchers today find it more realistic to present sonification as an important add-on to the range of data representation tools already available. Better still, they would like to collaborate not only with experts who are already used to listening, such as physicians, but also with domain scientists who commonly work with digital visualization techniques. By creating sonification plug-ins for a widely used data visualization software package, for instance, they enable domain experts to rely on their existing expertise and skills rather than being forced to familiarize themselves with the software used by sonification designers (Supper 2012a: 256; 2015a: 453).

In fact, when presenting their own work at conferences, members of the sonification community do not eschew diagrams, photos, or screen shots. And although the organizers of each International Conference on Auditory Display ask for sound examples, a recurrent complaint is that these are little used even in this heartland of sonification, and then mainly by the most experienced sonification proponents. This is partially due to the high likelihood of errors when using sound, fear of criticism regarding sound quality, or bad listening conditions in the venues. Instead, "data karaoke"—a phenomenon mentioned in Chapter 2—is often used to embody, highlight, illustrate, and authorize data and/or sonifications or to integrate them with the sonic form of the spoken presentation itself (Supper 2015b).

When claiming a place in the academic sun for sonification, Supper (2012a) has argued, the sonification community engages in "boundary work" as members shift rhetorically between connecting their work with and distancing it from visualization, depending on the contexts in which sonification is being propounded. Supper draws on STS findings on the ways that academic fields construct their cultural authority by demarcating themselves from non-science such as politics and religion (Gieryn 1995), crossing the borders of other disciplines (Klein 1996; Burri 2008), defining the conditions under which existing boundaries can be crossed (Halffman 2003), or showing how an emerging field differs from and resembles established disciplines (Amsterdamska 2005). As we will see in the next chapter, sonification promoters have used similar strategies when positioning sonification in relation to art and music.

Supper also stresses, however, that the "current situation in sonification echoes, rather than solves" the problem of finding information in large datasets (Supper 2015a: 458). As she has amply documented, most work in sonification focuses on tools and designs for sonification rather than the analysis of data. Critics within the community complain that most of their colleagues seem to think these information patterns simply "jump out" from the data once the sonification has been made. Again, the sonification community's preferred choice of historical examples is canny. The Geiger counter displays levels of a well-known scientific phenomenon, while "the audification of earthquakes and nuclear detonations rests upon the principle of pattern recognitions in instances where the patterns themselves are well understood" (Supper 2015a: 456). For sonifications of the big data sets that sonification researchers have in mind, in contrast, the patterns remain to be found and understood, and do not make themselves apparent automatically.

Moreover, sustained collaboration and shared listening would be necessary for the datasets to be properly understood by sonification specialists and the sounds to be properly understood by the domain scientists. This is still rare (Supper 2015a: 457). Partially as a result, the sonification community has not yet found its "killer application" for diagnostic listening. In informatics, the term killer application "refers to an application program so useful that users are willing to buy the hardware it runs on, just to have that program" (Juolo 2008: 76, cited in Supper 2012a: 255).

So far, and despite the success of the EEG sonifications, no sonification application has attracted enough interest for domain scientists to buy into sonification on that scale—a situation deeply deplored by sonification researchers.

CONCLUSIONS

This chapter has explained the conditions under which embodied listening, notably for purposes of "diagnostic listening," to mechanically registered auditory phenomena was replaced by the mechanical analysis and visual display of data as the preferred sensory mode for producing knowledge. In the 1920s and 1930s, enthusiasm for mechanical sound recording as the epitome of mechanical objectivity reached a peak in many scientific fields. But as a solution to problems of sensory subjectivity, this enthusiasm for—or epistemological value projected onto—sound recording proved rather short-lived. The spectrograph promised to sidestep the cumbersome and controversial manual notation of sound, and enabled easier integration with texts and other visualizations on paper. In many medical fields, the tape recorder never really took off, as other auditory and visual instruments of diagnostics had already nested in medical practice.

In both the heydays of mechanical sound recording and the spectrograph, new instruments acquired scientific authority by being calibrated according to ways of listening or visual modalities that had been used in earlier periods. The sensory calibration of sound in vision, or vice versa, could be material in character, requiring an alignment with tangible sources captured in the same or other sensory modes, or metaphorical, suggesting similarities with modalities employed in the past: sound spectrograms conceptualized as a microscopy of sound, for instance. For some ornithologists, spectrography could not convey scientifically relevant findings without manual inventions and without anchoring the visualizations in comparative analytic listening. It was feared that particular phenomena, such as phrases in bird song, would be lost if the visualization of data through sound spectrograms was not enriched by professional audition. After all, a scientist was more than just a registrant. Trained judgment was those ornithologists' ideal, complementing the notion of mechanical objectivity. At stake here was professional audition's authority to have a say in the establishment of knowledge. That was also the goal some researchers in the sonification community had in mind. They underlined the similarities between trained judgment in sonification and visualization. As yet, however, they have failed to gain significant ground for collaborative work on sonification in the world of science at large, perhaps because they also flag the differences between sonification and visualization.

In medicine and car mechanics, the move was not so much from mechanical sound recording to mechanical visual recording as from stethoscopic listening to meter reading. In US automotive mechanics, this happened much earlier than in Germany, a discrepancy that can only be understood by considering the changing relations of trust between car mechanics and motorists. While the state-backed crafts system kept German car mechanics the authorized and reliable owners of diagnostic listening, the social setting in the United States in the long run made meter reading the only way to assure customer trust in car mechanics. In Germany, the same shift was completed only in the late 1970s, with rationalization and the electrification of the car as arguments. Even then, the use of sound and listening for knowledge acquisition continued to crop up, or "to pop up" in the sciences. What should we make of this seemingly stubborn reappearances?

Notes

1. See "Bird organ" under "Bird instruments" on Oxford Music Online. The first serinettes were built in seventeenth-century France (*serin* is a French word for canary), but the instruments were most popular in the eighteenth and nineteenth century, also outside France. The Oxford Music Online entry claims that serinettes were used to "encourage" canaries to sing. In 1852, however, the French bird breeder Jules Jannin explained that serinettes could also be employed to teach caged canaries to sing particular songs (Jannin 1852: 27).
2. I would like to thank Cyrus Mody for suggesting this line of thought.
3. An oscillograph is "a device that generate[s] visual displays of electrical signals" (Thompson 2002: 96). It can be used to indicate, in waveform, any quantity that can be converted into electric energy. One such quantity is acoustic energy.
4. At that time, a sound spectrograph deciphered the sound signals of a magnetic sound recording by measuring the sound energy of particular frequency ranges in that signal with a frequency band filter. A stylus recorded the sound energy in each of the frequency bands on a "revolving roll of electrically sensitive paper" (Bruyninckx 2013: 123). For spectrographic

108   K. BIJSTERVELD

images, see http://exhibition.sonicskills.org/exhibition/booth4/graph-ical-notation-the-spectrograph/ and http://exhibition.sonicskills.org/exhibition/booth4/notating-bird-song-and-sound/ (both last accessed August 14, 2017). On the second link, scroll down to "Transcriptions of chaffinch song," last page.
5. In 1958, the Laboratory of Ornithology at Cornell University, the institute behind the Cornell Library of Natural Sounds, earned 10,000 dollars from royalties. The institute published its own record series and sound books: natural history books with aural illustrations on gramophone. It also sold its recordings as sound effects to the entertainment industry, including Disney and Warner Bros, and to businesses, which played the sounds of natural predators in pest control, for example (Bruyninckx 2015: 353).
6. Robert Bosch Company Archives, "Die ganze Werkstatt-Ausrüstung Bosch," 1969, File Number EF 001/009.
7. Michael Lynch (2013: 68) refers to a 1985 work by historian of medicine Christopher Lawrence showing that some late nineteenth-century Victorian gentlemen-doctors even considered the stethoscope to threaten the standing of their "incommunicable knowledge," because it might open the medical profession to specialists beyond their own elite ranks.

REFERENCES

Amsterdamska, O. (2005). Demarcating Epidemiology. *Science, Technology and Human Values, 30*(1), 17–51.
Brady, E. (1999). *A Spiral Way: How the Phonograph Changed Ethnography.* Jackson: University Press of Mississippi.
Brăiloiu, C. (1970/1931). Outline of a Method of Musical Folklore. *Ethnomusicology, 14*(3), 389–417.
Broeders, A. P. A. (2002). Het herkennen van stemmen. In P. J. van Koppen, et al. (Eds.), *Het recht van binnen: psychologie van het recht* (pp. 573–596). Deventer: Kluwer.
Bronfman, A. (2016). Biography of a Sonic Archive. *Hispanic American Historical Review, 96*(2), 225–231.
Bruyninckx, J. (2013). *Sound Science: Recording and Listening in the Biology of Bird Song, 1880–1980* (Ph.D. thesis, Maastricht University).
Bruyninckx, J. (2015). Trading Twitter: Amateur Recorders and Economies of Scientific Exchange at the Cornell Library of Natural Sounds. *Social Studies of Science, 45*(3), 344–370.
Bruyninckx, J. (2018). *Listening in the Field: Recording and the Science of Birdsong.* Cambridge: MIT Press.
Burri, R. V. (2008). Doing Distinctions: Boundary Work and Symbolic Capital in Radiology. *Social Studies of Science, 38*(1), 35–62.

Burri, R. V., Schubert, C., & Strübing, J. (2011). Introduction: The Five Senses of Science. *Science, Technology & Innovation Studies, 7*(1), 3–7.

Collins, H. M. (2013). Building an Antenna for Tacit Knowledge. In L. Soler, S. D. Zwart, & R. Catinaud (Eds.), Tacit and Explicit Knowledge: Harry Collins's Framework. *Philosophia Scientiae, 17*(3), 25–39.

Daston, L., & Galison, P. (1992). The Image of Objectivity. *Representations, 10*(40), 81–128.

Daston, L., & Galison, P. (2007). *Objectivity.* New York: Zone Books.

Fehr, J. (2000). "Visible Speech" and Linguistic Insight. In H. Nowotny & M. Weiss (Eds.), *Shifting Boundaries of the Real: Making the Invisible Visible* (pp. 31–47). Zürich: VDF Hochschulverlag AG an der ETH Zürich.

Gieryn, T. F. (1995). Boundaries of Science. In S. Jasanoff, G. E. Markle, J. C. Petersen, & T. J. Pinch (Eds.), *Handbook of Science and Technology Studies* (pp. 393–443). Thousand Oaks, CA: Sage.

Halffman, W. (2003). Boundaries of Regulatory Science: Eco/Toxicology and Aquatic Hazards of Chemicals in the US, England, and the Netherlands (Ph.D. thesis, University of Amsterdam).

Jannin, J. (1852). *L'Art d'élever et de multiplier les serins canaris et hollandais.* Paris: Chez Tissot.

Klein, J. T. (1996). *Crossing Boundaries: Knowledge, Disciplinarities, and Interdisciplinarities.* Charlottesville: University Press of Virginia.

Korpáš, J., Sadloňová, J., & Vrabec, M. (1996). Analysis of the Cough Sound: An Overview. *Pulmonary Pharmacology, 9*(5/6), 261–268.

Krebs, S. (2014). "Dial Gauge versus Sense 1-0": German Car Mechanics and the Introduction of New Diagnostic Equipment, 1950–1980. *Technology and Culture, 55*(2), 354–389.

Krebs, S., & Van Drie, M. (2014). The Art of Stethoscope Use: Diagnostic Listening Practices of Medical Physicians and "Auto Doctors". *ICON: Journal of the International Committee for the History of Technology, 20*(2), 92–114.

Latour, B. (1986). Visualisation and Cognition: Thinking with Eyes and Hands. *Knowledge and Society: Studies in the Sociology of Culture Past and Present, 6,* 1–40.

Lynch, M. (2013). At the Margins of Tacit Knowledge. In L. Soler, S. D. Zwart, & R. Catinaud (Eds.), Tacit and Explicit Knowledge: Harry Collins's Framework. *Philosophia Scientiae, 17*(3), 55–73.

Mills, M. (2010). Deaf Jam: From Inscription to Reproduction to Information. *Social Text, 102, 28*(1), 35–58.

Mundy, R. (2018). *Animal Musicalities: Birds, Beasts, and Evolutionary Listening.* Middletown, CO: Wesleyan University Press.

Potter, R. K., Kopp, G. A., & Green, H. C. (1947). *Visible Speech.* New York: D. van Nostrand Company.

Prescatello, A. R. (1992). *Charles Seeger: A Life in American Music*. Pittsburgh and London: University of Pittsburgh Press.

Sterne, J. (2003). *The Audible Past: Cultural Origins of Sound Reproduction*. Durham: Duke University Press.

Supper, A. (2012a). The Search for the "Killer Application": Drawing the Boundaries Around the Sonification of Scientific Data. In T. Pinch & K. Bijsterveld (Eds.), *The Oxford Handbook of Sound Studies* (pp. 127–150). Oxford: Oxford University Press.

Supper, A. (2012b). Lobbying for the Ear: The Public Fascination with and Academic Legitimacy of the Sonification of Scientific Data (Ph.D. thesis, Maastricht University).

Supper, A. (2012c). "Trained Ears" and "Correlation Coefficients": A Social Science Perspective on Sonification. In M. A. Nees, B. W. Walker, & J. Freeman (Eds.), *Proceedings of the 18th International Conference on Auditory Display, Atlanta, GA, USA, June 18–21* (pp. 29–35).

Supper, A. (2015a). Sound Information: Sonification in the Age of Complex Data and Digital Audio. *Information & Culture: A Journal of History, 50*(4), 441–464.

Supper, A. (2015b). Data Karaoke: Sensory and Bodily Skills in Conference Presentations. *Science as Culture, 24*(4), 436–457.

Supper, A. (2016). Lobbying for the Ear, Listening with the Whole Body: The (Anti-)Visual Culture of Sonification. *Sound Studies: An Interdisciplinary Journal, 2*(1), 69–80.

Thompson, E. (2002). *The Soundscape of Modernity: Architectural Acoustics 1900–1933*. Cambridge: MIT Press.

Van Drie, M. (2013). Training the Auscultative Ear: Medical Textbooks and Teaching Tapes (1950–2010). *The Senses and Society, 8*(2), 165–191.

Volmar, A. (2013). Listening to the Cold War: The Nuclear Test Ban Negotiations, Seismology, and Pyschoacoustics, 1958–1963. In A. Hui, J. Kursell, & M. Jackson (Eds.), Music, Sound and the Laboratory from 1750–1980. *Osiris, 28*, 80–102.

Willaert, T. (2016). De fonograaf en de grammofoon in de Nederlandstalige literatuur 1877–1935: Een media-archeologisch onderzoek (Ph.D. thesis, Leuven University).

Woolf, C. R., & Rosenberg, A. (1964). Objective Assessment of Cough Suppressants Under Clinical Conditions Using a Tape Recorder System. *Thorax, 19*(2), 125–130.

CHAPTER 5

Popping Up: The Continual Return of Sound and Listening

Abstract This chapter aims to explain why practices of listening continue to "pop up" as routes into knowledge-making despite the dominance of visualization in the sciences. It identifies three recent trends behind this phenomenon: the rise and versatility of digital technologies, the significance of somatic vigilance and synchronization in today's large instrument-based laboratories, and the role of the auditory sublime in the public fascination with sonification.

Keywords Versatility of digital technologies · Somatic vigilance · Synchronization · Auditory sublime

Hearing Gravitational Waves

On February 11, 2016, the *New York Times* published an online video with the attention-grabbing title "Ligo Hears Gravitational Waves Einstein Predicted."[1] The video was embedded in a news article announcing that a group of scientists behind LIGO, the Laser Interferometer Gravitational-Wave Observatory,

> had heard and recorded the sound of two black holes colliding a billion light-years away, a fleeting chirp that fulfilled the last prediction of Einstein's general theory of relativity. That faint rising tone, physicists say,

© The Author(s) 2019
K. Bijsterveld, *Sonic Skills*,
https://doi.org/10.1057/978-1-137-59829-5_5

is the first direct evidence of gravitational waves, the ripples in the fabric of space-time that Einstein predicted a century ago.[2]

The original press release about the gravitational waves, by the National Science Foundation, explained that during the collision, "a portion of the combined black holes' mass" had been converted to energy "according to Einstein's formula $E = mc^2$ "and had been emitted as a "strong burst of gravitational waves." LIGO had observed this by sonifying the measurements of the arrival time of laser light split into two beams, each reflected by one of two mirrors at the end of the arms of LIGO's L-shaped interferometer. A small time lag between the arrival time of the light beams, sonified in terms of frequency, expressed "the tiny disturbances the waves make to space and time as they pass through the earth."[3] Or, in the sonically rich words of David Reitze, LIGO Lab Executive Director at Caltech, during the press conference:

> Now, what LIGO does is that it actually takes these vibrations in space-time, these ripples in space-time, and it records them on a photo-detector, and you can actually hear them. ... It is the first time the universe has spoken to us through gravitational waves. And this is remarkable. Up to now we have been deaf to gravitational waves, but today we are able to hear them. That is just amazing to me.[4]

It was thus the sonification of visualized measurements that gave the *New York Times* as well as LIGO itself reason to talk about *hearing* evidence of a phenomenon journalists and scientists alike considered fundamental to our understanding of nature. A rising tone signaled that Einstein had been right.[5]

As this chapter will show, the references to sound in the publicity on gravitational waves is a fairly typical, if spectacular, example of how listening continues to "pop up" as a strategy for acquiring knowledge despite the shaky epistemological authority of sonic skills in the sciences. Given its contestation, what makes listening, and notably for the purposes of monitory and exploratory listening in our taxonomy, a felt necessity or appealing feature of the sciences? In this chapter, I answer that question by relating the recurrent return of sound and listening in the sciences to three issues: the rise of digital sound technologies and the portability and versatility of these tools; the need for somatic vigilance in industrial settings, operating theaters, and laboratories; and the construction by both scientists and artists of a public fascination with the auditory sublime.

THE DIGITAL: PORTABLE AND VERSATILE
SOUND TECHNOLOGIES

I have explained in the previous chapters how the portability of sound recording instruments affected what could be recorded—birds in fields accessible to trucks, for instance. It also affected who could be involved; thus, the rise of magnetic tape recorders for consumer use in the 1950s enabled ornithologists to create a moral economy of exchange with amateurs. With the rise of digital technologies, sound recording's portability and versatility acquired profoundly new meanings. Unprecedented levels of virtuosity could be attained in such matters as switching between analytical, synthetic, and interactive listening.

In ornithology, field research is a new experience now that portable digital databases of bird sounds on iPods and iPads enable on-the-spot comparison between what has just been heard and what can be found on the database. At times, this new option leads to false reports of bird observations, when ornithologists or amateur bird spotters assume they have heard a bird singing whereas in fact it was just a digital sound device playing the recording of bird vocalization. The technology does, though, allow recorded bird calls to be used to attract individuals of the same species to a particular spot during fieldwork, or prompt competing male birds to call in response to the taped ones. As a form of explorative interactive listening, playing recorded bird sound to elicit vocalizations had already taken off in the age of the magnetic tape recorder, but the larger numbers of recordings available to ornithologists today have changed the game. Moreover, portable computer devices and free audio imaging software also help birdwatchers to, as one ornithologist put it, see what they hear and hear what they see synesthetically, indoors and outdoors (Bruyninckx 2013: 167).

Alexandra Supper (2012, 2015: 451ff) has sketched three ways in which the rapid expansion of sonification initiatives since the early twenty-first century has been assisted by digital technologies. First, the digital age tremendously increased the options for sharing and circulating sound files, and thus for carrying out sonification. Earlier sonification enthusiasts had used flexi discs in the 1970s, or compact discs in the 1990s, as appendices to paper publications. The introduction of digital audio storage, and notably the MP3 format in 1996, made the distribution of recorded sound faster, less costly, and almost effortless, although MP3 files may, in principle, be protected against free distribution by means of digital rights management. The "preservation paradox" of digital technology—which enables the easy transfer and storage of digital audio, yet undermines prospects of long-term retrieval due to the rapid introduction of new formats (Sterne 2009: 64–65)—is a potential threat to the robustness of sound-for-knowledge, just as it is for visual digital information. But the fact that MP3 files can be inserted into electronic publications and integrated with texts and images, rather than having to be attached as addenda, has enhanced the epistemological credibility of sound. The advantages of inscription as set out by Latour, such as superimposition, are no longer restricted to the visual representation of data: "synaural" presentation is now possible as well—it has already been flagged in previous chapters. And as Florian Dombois, one of Supper's interviewees, put it: "A sound has to be published in order to count as an academic argument" (cited in Supper 2015: 452).

Second, the rise of digital tools for processing and creating sound, notably sound synthesis tools such as SuperCollider, MaxMSP, and Pure Data, have extended the possibilities for flexibly tweaking the parameters of sonification. Rather than simply transposing time-series data to waveforms in the human auditory range (that is, "audification"), sound synthesis allows for "much more complex mappings between data and sound and for many more audio parameters (such as pitch, timbre, duration, brightness and panning)" (Supper 2015: 452). Because many of these tools have their origins in the world of computer music, they also make the sonifications more aesthetically accessible. The downside of their roots in music, however, is that the tools were not originally designed to handle large data sets, the facilities for which have to be added and kept compatible with changing software packages.

This is all the more important given that, third, growing academic interest in and availability of big data have added urgency to the quest for sonification. A long-standing promise of sonification has been to create order in the chaos and abundance of data. To be sure, this does not mean sonification offers straightforward solutions. As discussed in the previous chapter, the sonification community is currently better equipped to give audibility to data that are already well understood than to extract new patterns and information from unfamiliar data (Supper 2015: 458). Nonetheless, the versatility of digital technologies has opened up many novel ideas and practices for the "transduction" (Helmreich 2012: 160) and "synesthetic conversion" (Mody 2012: 225) of signals from one sensory mode into another, thus bridging older divides between distinct disciplines and social domains. I will return to these forms of bridging in this essay's last chapter.

SOMATIC VIGILANCE: ATTUNING TO INSTRUMENTS AND TIME

Our case studies offer ample examples of listening for monitory purposes to the sound of machines in factories or to research instruments in laboratories. In twentieth-century industrial settings, workers considered the auditory surveillance of machines so crucial to their performance that they even hesitated to accept hearing protection (Bijsterveld 2008, 2012). Nowadays, such protection is compulsory in many countries, and most industrial machines are monitored using computer screens. In certain situations, though, operators may still listen to machines as auditory monitoring. An example comes from the observations and interviews carried out by Stefan Krebs in 2013 at Frogmore Mill in Hemel Hempstead, UK. Frogmore is now a heritage institution, but until 2000 it was the world's oldest mechanized paper mill still in operation. One of the operators referred to his experience on occasions when

> the noise was so great, that, you could, if you were tired, as you often were late at night, the noise, you would begin to hear things, so you begin to hear choirs or orchestral music, that kind of thing, just, just a kind of dream or an auditory daydream would come about, and it's something I actually found that I can control, so I could actually hear pieces of music that I knew well … and clearly what was going on was my brain … filtering out what it didn't need, and it wasn't the same as in a quiet room imagining the music, in that noise I was actually hearing it. (Operator cited in Krebs 2017: 43)

Interestingly, the operator additionally said that when he wandered off into auditory daydreams, he did not stop deploying his listening skills to monitor the machine's functioning. In fact, whenever his musical experiences were interrupted, he would know that something significant had changed. Apparently, he first transformed machinery noise into music; then, the musical patterns or breaches in those patterns informed him of the proper or problematic functioning of the machine. As Joy Parr showed in *Sensing Changes*, such rhythms may be deeply ingrained in people's corporeal experiences of the locations they inhabit (Parr 2010, 2015: 18).

Whereas Stefan Krebs interviewed paper-mill operators about their memories of sensory skills in the recent past, sociologist Sarah Maslen (2015) interviewed fifteen doctors from different disciplines about their listening experiences in the present day. In one of these interviews, an orthopedic surgeon explained that arthritis involves the loss of "the layer of cartilage that allow[s] for frictionless movement" in the joints. This is not visible on X-ray, but announces itself through "creaks" and "grates," sounding "like wheels that need oil." These and other bodily sounds are also relevant during orthopedic surgery. When surgeons are drilling bones to insert implants, for instance, changes in pitch tell them they have reached hollow or outer areas of the bones, helping them to navigate through bodies during operations (Maslen 2015: 61–62). This form of monitory listening enables surgeons to distinguish spatially between right and wrong: Yes, I need to be here, or No, it's the wrong spot.

In principle, this monitory quality also holds for navigation sounds "that are artificially produced and played back through magnetic speakers or piezoelectric units in medical equipment to indicate surgical operative tasks" (Schneider 2008: 2). The value of alarm signals generated by medical instruments in operation theaters and intensive care is felt to be less evident, however. Although auditory signals such as buzzers, beeps, sirens, pulses, or chimes in theory call the staff's attention to problems in the patient's condition or procedural faults, it is by no means clear whether alarm sounds actually enhance performance in hospitals and similar settings. The answer appears to depend not only on the character of the sounds, but also on the type of work and the workload of the people who must respond to the alerts. If staff have a high visual workload, for instance, auditory alarms seem to be useful (Edworthy and Hards 1999: 604), but when the overall workload is too high, operators may start to rely too strongly on alarms—whether visual or auditory (Endsley and Jones 2012: 155–157).

The proliferation of auditory medical alarms since the 1980s, partly due to a perceived need to protect the liability of the medical instrument manufacturers, has complicated the alarms' use. Anesthesia machines, artificial ventilators, blood warmers, electrosurgical units, hyperthermia systems, infusion pumps, monitoring systems, and pulse oximeters all have built-in alerts. Some of these may be masked by the cacophony of sounds, and even when the alarms are noticed, it may not be easy for medical staff to identify their sources or interpret their urgency correctly. Confusingly, different manufacturers offer different alarm sounds for the same variables, while different types of devices may generate similar sound alarms depending on their make.

A 2006 article discussing the interpretation of thirteen medical alarm signals by clinical engineers with different levels of experience concluded that the overall recognition rate was a mere 48 percent (Takeuchi et al. 2006). The International Organization for Standardization issued several recommendations on standardizing alarm sounds in the 1990s, but real standardization has failed to materialize, inspiring scientists to design aids such as an alarm sound database and a simulation set-up for training operating room attendants. The simulation enables users to listen to alarms in the context of an artificial hospital soundscape featuring speech, doctors' beepers, automatic doors, and music (Takeuchi et al. 2006; Schneider 2008). Auditory navigation and auditory surveillance are thus still significant monitory listening skills in hospital, though ones endangered by an over-abundance of alarms.

In fact, the large number of alerts may also elicit new sonic skills. Chapter 3 discussed experienced nurses tightening intensive care unit alarms to reduce the overload of alerts: an example of interactive monitory listening. Patients have less control, however, and Tom Rice reports that his "patient interlocutors often experienced the wards as being disturbingly noisy," alarms being one of the sources of such noise (2013: 29). Several of the nurses Anna Harris talked to during her fieldwork at an Australian hospital nostalgically evoked the relative tranquility of the intensive care unit (ICU) of the past. She cites one of them:

> Now there is an alarm for everything and they are forever tightening the alarms. The noise is horrific now. ... It's changed so much. There is no respite any more. [phone rings nearby] There is a sound for everything—to get in a door and another click when you leave. The [hand cleanser] dispenser makes a noise too! I remember the sound of billows in the ICU—it was quite peaceful, like white noise ... I could go to sleep to that noise. Gone are the days of peaceful ICU. (Field notes Anna Harris, Melbourne, October 21, 2013, cited in Harris 2015: 25)

It is worthwhile reflecting further on one of the reasons for this plethora of alarms, medical instrument manufacturers' fear of liability for non-functioning instruments. This implies an important new context for the epistemic relevance of sound: the alarms indicate both the experts' dependence on black-boxed machines—as manufacturers set the alarms—and, in some contexts, the need they feel to constantly monitor and discuss the machines' performance. Joeri Bruyninckx has shown the significance of these phenomena for sound and listening in modern science labs. In recent years, lab experiments have increasingly been organized around automated tools and expensive instruments that function as platforms for large numbers of researchers from different fields. Bruyninckx studied the handling of automated experimental protocols, carrying out extended ethnographic observations of and interviews with researchers and technicians in a Dutch lab for surface science, plasma science, and materials science (anonymized as PlasmaLab)[6]; he also examined user practices concerning the same type of platforms in three US labs for five months. The American labs worked with mass spectroscopic and nuclear magnetic resonance (NMR) techniques, for crystallographic characterization and the definition of molecular structures.

At first glance, it might be expected that using instruments with commodified software reduces reliance on researchers' sensory skills for monitoring the instruments. Indeed, programmed commands have been introduced to boost productivity and efficiency, and to standardize the experimental set-ups and enhance replicability. Bruyninckx (2018: n.p.) has shown, however, that these intentions do not mean the instruments are always or entirely trusted—such trust needs to be actively constructed and constantly reaffirmed. In hospital operating theaters, responsibility for the proper functioning of equipment seems to be delegated to the instrument manufacturers and the alarms, but the situation in experimental research labs is different.

Several of the researchers at PlasmaLab, for instance, had extensive experience with custom-built instruments, making them very aware of the effects of in-built parameters on the experimental results and keen to "open the hood" of ready-made tools, for instance by contacting manufacturers. Even without such experience, many of the researchers observed and interviewed considered the "knowability" of instruments key to assessing the set-up's stability and the reliability of experimental outcomes. "Sometimes," one doctoral researcher noted, "you actually think that the reactor has a personality" (Bruyninckx 2018: 11). For him and many of his colleagues, this means being aware of the instruments' whims in order to grasp unexpected outputs or breakdowns and to decide whether an experiment has succeeded or not. Understanding the internal working of instruments additionally contributes to researchers' independence from technicians, which in turn helps to build up their trust in their own and their peers' qualities as experimentalists—trust that also arises from the ability to answer critical queries about data in departmental meetings, for instance.

In pursuit of knowability, researchers often want to stand next to the instruments that provide their samples, hoping to materially witness the instruments' functioning on the spot. These practices embody "somatic vigilance," a "guarded attentiveness towards the technical conditions under which data are produced and interpreted." Somatic vigilance is more than the organized skepticism considered typical of science: it is a "tactic used by researchers to calibrate trust judgments" within the material, social, and knowledge regimes of their research settings (Bruyninckx 2018: 3, 7). Bruyninckx illustrates it with sensory examples of monitoring. These include reading graphs and numbers indicating the instruments' output, but also touching the instruments to check for heat or vibrations and listening to their sounds:

> The setup is automated so that it can be operated fully via the desktop monitor, but I always listen. You know that when you enter this [value], you should hear this sound ... . I don't trust the button {pause} you know, it is just a machine, something can go wrong. When I hear it, I know it for certain. (Field notes, 11 July, 2013, cited in Bruyninckx 2018: 18)

Similarly, the operational rhythm of the lab's entire soundscape tells researchers whether experiments done by others are running smoothly or signal unsafe situations. These examples show once again that the purpose of monitory listening can call for both synthetic listening (to all audible sounds at the same time) and analytic listening (focusing on one or a few sounds amidst everything that is audible).

Somatic vigilance is not limited to science researchers. During his fieldwork at the American chemistry and biology labs, Bruyninckx closely followed technicians in their day-to-day work, and happened upon the following instruction note near one of the instruments:

> Attention all Bruker 600 Users!!!
> If you do not hear the cryoprobe's helium pump
> "chirping", <u>DO NOT</u> use the instrument!
> STOP
> This means the probe is not working properly
> And you will NOT get a spectrum.
> Thanks.[7]

In this as in the other labs, technicians are responsible for the smooth operation of the machines and systems that form the heart of the workflow. Bruyninckx noticed that as they fulfilled this responsibility, technicians commonly rely on their experience of what research instruments "should look, feel, smell, and sound like" under normal circumstances, recalling the somatic vigilance of the plasma researchers just discussed. Some technicians not only acquire their own situated and embodied skills, but also train the user-researchers by calling their attention to these sensory specifics, among other things warning them that relevant sounds may be masked by the noise of other laboratory instruments. User instructions to "'listen for a click,' 'wait for the pzzzz,' or ensure that no 'hiss' or 'chirping' can be heard" aim to persuade users to monitor the instruments' functioning, but they also, or especially, encourage responsiveness to the rhythms of the machines more generally. They help the technicians to synchronize "users' temporal expectations with their instruments' rhythms by redirecting their attention, inviting them to open their bodies and allow themselves to be temporarily affected by an instrument in use" (Bruyninckx 2017: 834).

Such synchronizations, Bruyninckx argues, are vital to today's lab culture. As large, shared, and expensive instruments such as NMR proliferate, their efficient and cost-effective use has become increasingly important. This means that the platform's "organizational time"—its temporal management—needs to be attuned to its "instrumental time." Bruyninckx distinguishes three forms of organizational time. "Scheduled time" refers to the time slots (for example: ten minutes during prime time) assigned to individual users or groups of researchers working with the platform technologies. In "billing time," these slots are translated into costs for particular departments by computer systems that track log-in and log-out shifts, while "strategic time" reflects management decisions on "long-term research activities, research lines, and instrumental acquisitions." Instrumental time, in contrast, alludes to the "sequences, rhythms, and durations in activities of repair, maintenance and operations" that are specific to particular research instruments and protocols (Bruyninckx 2017: 828–830).

The work done by technicians is crucial for aligning the three forms of organizational time with instrumental time. When the replacement of a machine is postponed in strategic time, for instance, wear and tear is likely to affect its performance, and therefore instrumental time. Technicians often play a vital part in tackling and resolving such slippages, and their instructions on attentive monitory listening, such as "wait for 3.3 minutes until the noises stop," are particularly important (Bruyninckx 2017: 832–833). These synchronizations and monitory alerts, together with technicians' prioritization of particular tasks and their repair work to prevent system breakdowns, are by no means phenomena at the margins of contemporary science—they are at its very heart.

Auditory Sublime: Promising Wonder and Awe Through Sound

"Popping up" is exactly what has been happening with the sonification of scientific data since the turn of the twenty-first century. Alexandra Supper had no problems at all gathering many recent cases in fields as diverse as the geosciences, neurology, high energy physics, genetics, astrophysics, and microbial ecology (Supper 2012, 2014, 2015).[8]

Some of these sonification projects have been initiated by artists. An example is the sound installation *The Place Where You Go To Listen*, created by composer John Luther Adams in 2006 and located at the Museum of the North, University of Alaska. Among its sounds are "sustained chords" that sonify data on the position of the sun, and "deep rumbles" sonifying registrations from several of Alaska's seismological stations (Supper 2012: 39–40). Other events have been organized by researchers, such as Gerold Baier and Thomas Hermann's sonification of the electroencephalogram of an epileptic seizure at the Wien Modern festival in 2008 (Supper 2014: 34–35). In a third group of sonifications, scientists and artists collaborate. For LHCSound, online since 2010, physicist Lily Asquith worked with software specialists, the musician Richard Dobson, the composer Archer Endrich, and others to sonify particle detection data, including data about the Higgs boson so famously reported in 2012 (Supper 2014: 40). The project's legacy can still be found on the website of the Large Hadron Collider (LHC), the world's largest and most powerful particle accelerator, at CERN, the European Organization for Nuclear Research. It features CERN scientists playing musical instruments such as the harp, clarinet, and violin as "LHChamber Music," while reading scores that are sonifications of their LHC data.[9] These and other examples often sound like contemporary classical compositions, ranging from mildly to wildly avantgarde.

When Supper talked to the scientists involved in sonification projects, however, many of them said that in their day-to-day work, understanding data through sound was actually less important than the media coverage of talks, concerts, festivals, and web events suggested. A case in point is sonification in asteroseismology, a subfield of astrophysics that aims to understand the internal structure of stars by observing their pulsations. These observations are relevant because the stars' variations in brightness are thought to result from oscillations in the ionization equilibrium in their outer layers. In turn, the oscillations and their frequency spectra are dependent on the stars' mass and radius. Oscillation modes can therefore give scientists information about the properties of the stars' cores, which are hard to study any other way. In lectures for students and talks for general audiences, astrophysics professor Conny Aerts frequently explains these phenomena using stellar sonifications: "synthesized, sped-up sounds based on the visual observations of stellar oscillations" (Supper 2012: 43). She has also collaborated with composer Willem Boogman, whose piece *Sternenrest* sonifies the data on one specific star and uses surround sound to position the audience right in the middle of those data. Yet Aerts emphasizes that she and her colleagues tend to study oscillations visually rather than sonically. The sonifications are almost exclusively employed to introduce students to astrophysics or reach out to the general public.

Why is it, though, that scientists find sonification so helpful in those communicative situations? And what motivates artists to use it? In the world of modern music, Supper explains by reference to musicologist Richard Taruskin, sonification responds to a twentieth-century trend to regard music as a canvas for the objective and the material rather than as the expression of individual, Romantic subjectivity. Adams, for instance, defines *The Place Where You Go To Listen* as art produced by natural phenomena. Against that background, it is understandable that artists often take the exact relationships between data and sound more seriously than scientists do when presenting sonifications to the public (Supper 2014: 39–40). Additionally, sonification promises to compensate for the loss of "deep structure" that audiences began to experience when electronic music departed from classical music (artist John Dunn cited in Supper 2014: 41)—or, perhaps, the painful void faced by many listeners to electronic music when it rendered conventional harmonic and rhythmic patterns obsolete.

For scientists, popular sonifications embody another promise: that of evoking an "auditory sublime" in those who listen (Supper 2012: 71, 2014: 34). Traditionally, the Kantian sublime stands for experiences of "infinity and unimaginable greatness" elicited by natural phenomena such as storms or mountains—observed at a safe distance, yet with an emotional ambivalence in which awe and pleasurable wonder are mixed (Supper 2014: 44–45). The notion of the sublime has since been applied to the experience of art, architecture, and grand technologies as well. Supper identifies it in the rhetorical, musical, technical, and spatial means by which scientists foster sonification in collaboration with artists. Auditory and musical metaphors abound in the texts accompanying those sonifications, recalling Yolande Harris's (2012) findings when she examined bioacoustics research on the underwater sounds of whales.

Natural phenomena, Supper shows, are said to "speak" to their audiences; they have something to "tell." The synthesis of proteins adheres to "a genetic score," stars have a "voice" and "sing," and humans can "eavesdrop on the brain" (Supper 2012: 54–57). References to the sublime are ubiquitous, in "the wonders of the cosmos, the dangers of the earth, the inconceivability of particles, the powers of genes and the complexity of the brain" (Supper 2014: 47).

Many sonification makers present sound as the perfect means to elicit sublime experiences and enable listeners to emotionally connect to the mysteries of nature. The three-dimensionality of sound is regarded as vital to these experiences—sound offers a particular sense of presence, immersion, and intimacy with natural phenomena, without actually getting dangerously close. Such immersion can be enhanced technologically, as with surround sound speakers. At the same time, the sonifications are intended to enthrall: the sounds can be loud and uncomfortable, but they may equally be "eerie" and "otherworldly," "chilling," and "disquieting" (Supper 2014: 47). Together, these dimensions offer virtual access to and deeper understanding of natural phenomena such as stars, volcanoes, and particles "that are too far away, too close-by, too big, too small, too high, or too low to be experienced in an unmediated way" (Supper 2012: 72). It is the exciting expectation of the sublime, of experiencing nature in its most overwhelming forms through sound, that scientists believe will attract the general public to large-scale research projects. Rather than inviting those audiences to listen diagnostically and analytically, I would add, the scientists seem to aim for experiences of exploratory listening in its synthetic mode.

CONCLUSIONS

When participants at LIGO Lab's press conference on gravitational waves said that the universe had "spoken" to us and that we could not remain "deaf" to the waves, this was clearly an instance of invoking the auditory sublime to attract the attention of the public. The LIGO scientists were trying to bring a complicated natural phenomenon closer to the wider audience while simultaneously inspiring a sense of respectful distance. In many natural science projects, an additional step has been collaboration with sound artists in order to help the public develop a fascination with the otherwise often intangible products of contemporary science.

The increasing versatility and portability of digital technologies is a very productive aspect of these processes. Not only does it assist the continual transduction of data from one sensory domain into another, from the visual into the auditory and vice versa; digital sound technologies also enable sound variables to be presented in ways that stage patterns in the data in more accessible forms than in the past. The rise of music software with easy-to-work-with interfaces has been instrumental in extending the options now available to sonification experts.

Increased digital versatility and promises of sublime experiences offer clues as to why displaying data in terms of sound continues to pop up and has even grown in importance despite the sonification community's failure to find a "killer application." And although sonification specialists still appeal for exploratory and diagnostic listening to data, monitory listening has gained increasing relevance in science labs—paying close attention to the rhythm of ever more expensive instruments can prevent them from running out of control and requiring costly repairs.

This chapter has also shown that attending to the role of sound allows us to articulate new developments in the sciences, such as the synchronization of work required in labs with large, grant-greedy set-ups or scientists' use of sonification in outreach activities. But the wider mechanisms behind the recent rise of the sonic versions of somatic vigilance and the exploitation of the auditory sublime have not yet been set out in detail. They form the topics of this essay's final chapter, on the relationship between listening for knowledge and issues of timing, trust, and accountability in the dynamics of science, technology, and society.

Notes

1. Dennis Overbye, Jonathan Corum, & Jason Drakeford, "Ligo Hears Gravitational Waves Einstein Predicted," Video *The New York Times Online*, February 11, 2016, at http://nyti.ms/1V6puGS (last accessed July 25, 2016).
2. Dennis Overbye (2016), "Gravitational Waves Detected, Confirming Einstein's Theory," *The New York Times Online*, February 11, 2016, at http://www.nytimes.com/2016/02/12/science/ligo-gravitational-waves-black-holes-einstein.html?_r=0 (last accessed July 25, 2016).
3. "Gravitational Waves Detected 100 Years After Einstein's Prediction," February 11, 2016, at https://www.ligo.caltech.edu/news/ligo20160211 (last accessed July 25, 2016).
4. National Science Foundation, "LIGO Detects Gravitational Waves—Announcement at Press Conference (Part 1)," at http://mediaassets.caltech.edu/gwave#conf, at 10'38 ff. (last accessed July 25, 2016).

5. Listen to the LIGO-edited sound files at https://www.youtube.com/watch?v=KzVDlFpaRRk&sns=em (last accessed July 25, 2016).
6. This particular case study had two phases, an explorative one (two months) by Aline Reichow in 2011, and a systematic phase executed by Joeri Bruyninckx (nearly three months) in 2013. Bruyninckx interviewed fifteen researchers and technicians.
7. Field notes Joeri Bruyninckx, facility A, January 30, 2014.
8. For a few examples, see http://exhibition.sonicskills.org/exhibition/booth1/how-are-sonifications-made/ and http://sss.sagepub.com/site/Podcasts/podcast_dir.xhtml (both last accessed at August 14, 2017).
9. http://home.cern/about/updates/2014/10/cern-scientists-perform-their-data (last accessed January 20, 2017).

REFERENCES

Bijsterveld, K. (2008). *Mechanical Sound: Technology, Culture and Public Problems of Noise in the Twentieth Century*. Cambridge: MIT Press.

Bijsterveld, K. (2012). Listening to Machines: Industrial Noise, Hearing Loss and the Cultural Meaning of Sound. In T. S. S. Reader (Ed.), *Jonathan Sterne* (pp. 152–167). New York: Routledge.

Bruyninckx, J. (2013). *Sound Science: Recording and Listening in the Biology of Bird Song, 1880–1980* (Ph.D. thesis, Maastricht University).

Bruyninckx, J. (2017). Synchronicity: Time, Technicians, Instruments, and Invisible Repair. *Science, Technology and Human Values, 42*(5), 822–847.

Bruyninckx, J. (2018). Instrument Trust and Somatic Vigilance in Experimental Physics. *Science as Culture*.

Edworthy, J., & Hards, R. (1999). Learning Auditory Warnings: The Effects of Sound Type, Verbal Labelling and Imagery on the Identification of Alarm Sounds. *International Journal of Industrial Ergonomics, 24*(6), 603–618.

Endsley, M. R., & Jones, D. G. (2012). *Designing for Situation Awareness: An Approach to User-Centered Design* (2nd ed.). Boca Raton, CA: CRC Press.

Harris, Y. (2012). Understanding Underwater: The Art and Science of Interpreting Whale Sounds. *Interference: A Journal of Audio Culture.* Available at http://www.interferencejournal.org/understanding-underwater/. Last accessed August 18, 2017.

Harris, A. (2015). Eliciting Sound Memories. *The Public Historian, 37*(4), 14–31.

Helmreich, S. (2012). Underwater Music: Tuning Composition to the Sounds of Science. In T. Pinch & K. Bijsterveld (Eds.), *The Oxford Handbook of Sound Studies* (pp. 151–175). Oxford: Oxford University Press.

Krebs, S. (2017). Memories of a Dying Industry: Sense and Identity in a British Paper Mill. *The Senses & Society, 12*(1), 35–52.

Maslen, S. (2015). Researching the Senses as Knowledge: A Case Study of Learning to Hear Medically. *The Senses & Society, 10*(1), 52–70.

Mody, C. C. M. (2012). Conversions: Sound and Sight, Military and Civilian. In T. Pinch & K. Bijsterveld (Eds.), *The Oxford Handbook of Sound Studies* (pp. 224–248). Oxford: Oxford University Press.

Parr, J. (2010). *Sensing Changes: Technologies, Environments, and the Everyday, 1953–2003.* Vancouver: UBC Press.

Parr, J. (2015). The Senses and the History of Technology. *Technikgeschichte, 82*(1), 11–25.

Rice, T. (2013). *Hearing and the Hospital: Sound, Listening, Knowledge and Experience.* Canon Pyon: Sean Kingston Publishing.

Schneider, M. (2008). Database Concept for Medical Auditory Alarms. In *Proceedings of the 14th International Conference on Auditory Display, Paris, France, June 24–27* (pp. 1–8).

Sterne, J. (2009). The Preservation Paradox in Digital Audio. In K. Bijsterveld & J. van Dijck (Eds.), *Sound Souvenirs: Audio Technologies, Memory and Cultural Practices* (pp. 55–65). Amsterdam: Amsterdam University Press.

Supper, A. (2012). *Lobbying for the Ear: The Public Fascination with and Academic Legitimacy of the Sonification of Scientific Data* (Ph.D. thesis, Maastricht University).

Supper, A. (2014). Sublime Frequencies: The Construction of Sublime Listening Experiences in the Sonification of Scientific Data. *Social Studies of Science, 44*(1), 34–58.

Supper, A. (2015). Sound Information: Sonification in the Age of Complex Data and Digital Audio. *Information & Culture: A Journal of History, 50*(4), 441–464.

Takeuchi, A., Hirose, M., Shinbo, T., Imai, M., Mamorita, N., & Ikeda, N. (2006). Development of an Alarm Sound Database and Simulator. *Journal of Clinical Monitoring and Computing, 20*(5), 317–327.

CHAPTER 6

Ensembles of Sonic Skills: Conclusions

Abstract This chapter combines a diachronic with a synchronic approach. It explains how different ensembles of sonic skills, or sets of sonic skills in specific settings, come to prevail with shifting relations between science and technology, science and the professions, and science and society. These ensembles reflect the significance of timing, trust, and accountability in the dynamics of science.

Keywords Ensembles of sonic skills · Science dynamics · Timing in the sciences · Trust in the sciences · Accountability in the sciences

INTRODUCTION

This final chapter builds on the previous ones by showing that listening modes and sonic skills in the sciences come in ensembles. By this, I mean that it is in particular configurations—in recurrent modes and with specific tools—that listening for knowledge has been considered useful or even vital to science, medicine, and engineering. "Ensembles of sonic skills," then, are not simply *sets* of sonic skills, but also the *settings* within which sonic skills are appropriated. These ensembles will be discussed here in terms of family resemblances: although the sets and settings may differ in detail, they have certain characteristics in common. The association with musical ensembles is no coincidence. In music, the terms string quartet or wind quintet do not only denote particular combinations of

© The Author(s) 2019   131
K. Bijsterveld, *Sonic Skills*,
https://doi.org/10.1057/978-1-137-59829-5_6

musical instruments, but also the repertoires they play. And ensembles are not likely to perform on any conceivable occasion—some occasions are more suitable than others. It is set and setting together that make the difference.

Thinking about ensembles will help us to understand, at a more structural level than in the previous chapters, where and why sonic skills survived or returned, as well as where and why sonic skills lost their relevance. Whereas the preceding chapters were either synchronic or diachronic in their approach, this one is synchronic in identifying ensembles of sonic skills, but diachronic in analyzing the scientific dynamics in which they operate. I aim to show how the changing relationships between science and technology, science and the professions, and science and society both enable and constrain the usefulness and legitimacy of listening for knowledge in science, medicine, and engineering. Such shifts are best captured by focusing on issues of timing, trust, and accountability. Scientists and professionals, the topics they examine, the skills and tools they use for engaging their ears, and the wider public all need to be taken into account if we are to understand what sound and listening "do" in the sciences.

In this undertaking, I also hope to slightly revise the historiography of the position of music in the sciences, and to complicate two existing accounts about the senses in the sciences: the claim that scientists sought out to listen whenever the subject under study was not directly accessible, and the emphasis on the significance of immutable mobile, visual inscriptions for the sciences. But let me first address the relevance of ensembles of sonic skills, and how these ensembles cut across the diversity of instances of listening described in the previous chapters.

Timing—And the Relations Between Science and Technology

A widely accepted, Latourian argument for the dominance of visualized presentations of data in science has been their immutability and their convenient combinability with texts on paper and screens, which contributes to the easy circulation of scientific claims. I have referred to this several times: scientific inscriptions can be synoptically organized and mapped onto each other. That option has also been opened to sound, in synaural form, by the rise of digital sound technologies. MP3 files are now easily combinable and distributable, helping to stimulate the rise of sonification.

This line of reasoning puts most of its faith in processes of publishing in the sciences, as it extends the networks of "acting at a distance." Yet oral presentations are still crucial elements of doing science, medicine, and engineering, whether at international conferences, in departmental seminars, or for teaching, with or without gestures in the air or the use of black- and whiteboards. Moreover, as we have discussed in Chapter 2, sound is highly compatible with oral presentations. Alexandra Supper (2015, 2016) made this point when she discussed the phenomenon of "data karaoke" in the sonification community. When presenting their data at conferences or meeting up with domain scientists, sonification specialists frequently sing their data as they speak about their findings. In doing so, they are embodying their data, often combining them with gesticulation and using the voice to highlight particular aspects of their results, and thus of their analysis. This is especially helpful when working with domain experts, who are usually less well versed in diagnostic listening to sonifications. As an additional advantage, users can trust their own voice rather than run the risk of sound files or speakers failing at the critical moment, for instance because they are not connected properly or otherwise out of order. Voices and gestures are easily available, any moment, anywhere. Similarly, other forms of embodied representation survived because it enabled practitioners to The choice of a particular sensory modality thus depends both on compatibility with the modalities already dominating the setting and on the importance of *timing* a particular action in that setting.

Equally, opinions as to whether or not particular tools, such as the phonograph or the magnetic tape recorder, are appropriate in a particular domain of science may be guided by issues of timing. In the world of naturalists sound-hunting for easily disturbed birds, for instance, the cumbersome combination of phonograph and wax heaters was not immediately taken up. Magnetic tape recorders, in contrast, embodied the kind of portability welcomed in the field. Even in the analysis of a visual inscription, the sound spectrogram (assumed to be efficient for swiftly surveying results), some believed an auditory approach could contribute to the analysis, for instance by creating crude classifications by ear before zooming in on the details. Again, in these situations, the need for careful timing of a particular action in the process of doing science informed the choice of one particular sonic instrument over another, or the decision to start with listening before focusing on visual inspection.

In this essay's second chapter, I also pointed out the significance of urgency—the imperative need to take action at a particular moment in time—for the survival of verbal expressions or manual notations of sound. When teaching medical students or tracking birds in the field, such ways of capturing sound may be the most efficient strategy for recording data or communicating knowledge on the spot. Similarly, the stethoscope survived in situations where high-tech instruments were not available or where urgency reigned, such as in wartime hospitals or countries unable to access the full range of expensive diagnostic instruments. Indeed, even in high-tech contexts stethoscopes may be of use in urgent situations, as a story recounted by European Space Agency cosmonaut André Kuipers illustrates. Kuipers, a Dutch physician by training, worked with a Russian aerospace engineer, the commander, and an American chemical engineer on board the International Space Station (ISS) for over six months in 2011–2012. One day, a space capsule arrived with new and essential supplies. Once the capsule had docked to the station, the commander made rotating movements to open the door. But whatever he tried, the door remained closed. He and his colleagues had no clue what might be causing the problem. It was in this setting that Kuipers' identity as a doctor came in useful, and led him to his stethoscope—in order to listen to the door. This enabled him to work out whether or not there was a mechanical issue. The clicks and sounds of rotating parts told him that the door itself was not the problem. It simply required the strength of three cosmonauts to open the door.[1] Kuipers's diagnostic listening skills saved the day in this expedition episode; in a context of urgency, the stethoscope turned out to be a highly efficient instrument.

These observations help me to qualify Sophia Roosth's (2009) suggestion that scientists tend to rely on their ears or on listening tools once they cannot directly access (in her examples: see) the phenomena of their interest. This is not the full story of why scientists come up with the idea of listening or are ready to explore it. First, prior listening experiences may help scientists to reconceptualize a problem or phenomenon as being open to listening, as the ISS story shows. In another example, the realization of sound's role in survival during World War I paved the way for academics to acknowledge the Geiger counter's usefulness in measuring radiation. Second, the societal relevance of a problem plus the availability of new instruments may also frame it as amenable to a listening approach. It has long been known that icebergs—one of the examples with which this essay began—can only be partially seen. What has opened up the underwater portions of icebergs more systematically to the epistemology of listening is the current sense of an urgent climate problem, in combination with the availability of high-quality hydrophones and ways of digitally transmitting sound files. Likewise, early nineteenth-century physicians already had access to percussion as a way of making the body legible; the readiness of many of them to embrace the stethoscope was spurred by a new desire to create physical distance between patient and doctor. Third, for a sound technology to be accepted as an instrument of investigation, it often requires material or conceptual sensory calibration with the sensory modalities of equipment used in the past. And fourth, as so many STS scholars have argued before us, making objects aurally (or visually) researchable redefines the character of those objects themselves—think of what the clean, visually informed sound captured by microphones with parabolic reflectors meant for definitions of and approaches to bird vocalization. All this complicates the idea that as soon as an existing phenomenon cannot be seen, the time has come to start listening.

Returning to the significance of timing one's actions for listening in the sciences, as an ensemble of sonic skills, we should also recall the work of synchronization as a structuring aspect of modern-day laboratories. For facility staff as well as scientists, showing an ability to adjust to the temporalities of planning schedules and instruments' workings expresses their efficiency to their superiors. In Joeri Bruyninckx's study, monitory listening to the instruments, and talking about sound when training newcomers to the lab, was an important dimension of an "embodied awareness of time"(Bruyninckx 2017: 840).

Trust—And the Relations Between the Sciences and the Professions

The extent to which audiences have invested trust, a solid belief in the reliability and responsibility of a person's or institution's actions, in those who listen for knowledge has deeply affected the levels of justification required for experts to rely on their ears. Such audiences ranged from patients and customer-drivers to scientific peers.

When, in the late 1920s, motorists started to question the expertise of the car mechanics who repaired their cars, the existing ensembles of sonic skills came under pressure. German mechanics responded both by casting doubt on the listening capacities of their clients and by creating a societally acknowledged jurisdiction for their own trade. They reclaimed the ability to listen diagnostically to engines as their exclusive skill—at the expense of drivers, who were to limit the use of their ears in automotive contexts to monitory listening. At the same time, they successfully established a guild-like system designed to strengthen clients' trust in the mechanic's expertise in diagnostic listening. In the United States, the same repair crisis played out rather differently. As US mechanics did not have Germany's tradition of protected trades, client trust in their sensory skills was easily undermined; that trust was transferred to measuring instruments featuring visual displays. Clients might be just as unable to read these instruments as to understand the mechanic's sonic skills, but the promise of transparency had the required effect.

The German mechanics modeled themselves on doctors, claiming the same type of tacit, clinical expertise and a similarly well-founded authority to describe sounds in their own words. They maintained this auditory autonomy until, with the rise of electronics, automotive technology itself had less to say through sound. In the world of medicine, the distance created by the stethoscope initially enhanced patients' trust in doctors. When medicine had to comply to the same standards of objectivity as science, it moved towards standardization, but at the bedside and in teaching contexts, doctors may still refer to sounds and their diagnostic meanings, in whatever ways they find effective.

Anna Harris has noted that despite the decreasing relevance of auscultation in hospitals and skepticism about its future, students worldwide are still being trained in both auscultation and percussion. This is because senior physicians and medical educators believe that teaching medical students "how to use a stethoscope to find heart murmurs or how to percuss the borders of the liver" is vital for honing their "sensory awareness" (Harris 2015). What is important here, Harris and doctor-educator Eleanora Flynn claim, is the education of students' attention—a notion introduced by James J. Gibson and taken up by Tim Ingold and Bruno Latour. In this "education of attention," learning to listen feeds into learning to pay attention, "a quality that is important for all sensory skills, not only those of auscultation and percussion" (Harris and Flynn 2017: 4). Students learn to notice differences in sensation, a skill many medical professionals consider crucial to their own and students' observatory and diagnostic abilities. Even if the skills of percussion and auscultation are on the decline as diagnostic techniques in hospital settings, therefore, training in these skills fulfills a "pedagogical purpose," thereby contributing to students' medical identities—not least through how they practice on, experiment with, and discover their own bodies (Harris 2016: 52).

In labs, sensory awareness is no less important, as an embodied vigilance concerning the material conditions of experiments helps to synchronize organizational and instrumental time. Whereas many doctors and mechanical engineers pride themselves on their sonic skills as much as other problem solving skills, lab technicians tend to make their skills invisible. The technicians observed by Bruyninckx mobilized "valuable social capital" to arrange replacements and free advice in order to keep the systems running. They did so largely out of sight of the lab's management, however, apparently fearing that technicians' visibility would actually highlight "ruptures in the operation they are charged (and paid) to maintain" (Bruyninckx 2017: 840–841). Keeping their skills out of view of those higher in the hierarchy contributes to an impression that everything is running smoothly, paradoxically reinforcing trust in technicians' work, which is not sanctioned by professional jurisdiction in the way achieved by mechanics and doctors. But it also makes the vital importance of monitory listening for contemporary lab life less obvious to STS researchers—unless one specifically focuses, as we have done, on the role of sound in science.

The inverse relationship between trust invested in experts and the justification required to employ sonic skills is perhaps best illustrated by those who are entrusted and hired by governments to display generalized distrust, such as the secret service, police, or military. In those settings, listening to the "enemy" and analyzing sound has never lost its importance, despite these activities having been backed up by visual techniques as well. In turn, those monitored—from soldiers, spies, and criminals to prisoners (Rice 2016)—have availed themselves of nonverbal self-expression and counter-eavesdropping on their sonic environments in order to stay under the radar of surveillance. Clearly, we should not forget that sonic skills are not limited to those working in science, medicine, and engineering (Bruyninckx and Supper 2016: 2), even if these groups have taken center stage in this essay.

Accountability—And the Relations Between Science and Society

Technicians may have various reasons for carrying out some of their repair work discreetly, "outside the organization's 'visible' range of auditing and accounting tools" such as the billed time of instrument time. They do so "to maintain their professional status, to protect their social capital, or in order not to provoke a further tightening of organizational temporalities" (Bruyninckx 2017: 841). This may, however, render inadequacies in organizational plans or replacement schemes invisible as well and sustain a high tempo of research, potentially to the point of compromising researchers' ability to properly understand, work with, and interpret the instrumental set-up. Synchronization work supported by the senses thus keeps in place the system of accountable time.

The experienced nurses who adjusted the alarms of intensive care instruments to prevent false alerts and unnecessary disruptions to their workflow similarly sidestepped the constraints of accountability. But they did so only temporarily; they also readjusted the alarms to more sensitive settings to ensure that less experienced staff would notice the slightest potential patient problem, and to protect the liability of the hospital and medical instrument manufacturers. In other words, tweaking alarms once again leaves the system of accountability intact. In the lab example, monitory listening in analytic and synthetic modes had to prevent listeners from noticing too little, whereas in the hospital, the senior nurses' interactive monitory listening had to prevent them from noticing too much. In both cases, the accountability structure triggered their use of sonic skills.

Accountability is also relevant for understanding the recent rise of sonification, and especially musical sonification, in the sciences. As explained in the previous chapter, sonification is considered a highly effective tool for attracting the attention of the general public with the promise of an auditory sublime. This close link between science and music seems to contradict the historiography of the evolution of music's position in the sciences. Alexandra Hui studied German-Austrian scholars and scientists examining the sensation of sound between 1840 and 1910. Their laboratory work, she argues, was initially "bound up with musical training and therefore musical aesthetics. Musical skill was a scientific skill" (Hui 2013: 145). Hermann Helmholtz's study of overtones, for instance, compared the workings of the inner ear to those of the piano: when striking a particular piano string, strings with corresponding harmonics would resound, and so too would the cortical fibers of the inner ear. Ernst Mach's research on the relationship between sensation and attention also depended on musical examples. He showed that listeners perceived the same sequence of chords quite differently when they focused on either the lower or the upper tones.

By the end of the nineteenth century, however, "the value of musical skill had become contested" (Hui 2013: 145). This was due to a rising interest in non-Western music and a changing musical aesthetics—such as Arnold Schönberg's twelve-tone music, which undermined the dominance of the Western music system—as well as to changing opinions about the ideal experimental listener. In a debate between Wilhelm Wundt and Carl Stumpf over just-noticeable pitch differences, for example, Wundt promoted the use of thousands of listeners, including musically untrained listeners, to substantiate statistically valid claims about sensation. In contrast, Stumpf believed that only listeners with musical expertise could hear the relevant distinctions; the outcomes of studies using musically untrained listeners were invalid. Wundt's approach won the day.

A similar trend seems to have been at work in ornithology as well, as it shifted from musical notation, to graphical notation, to the spectrographic rendering of sound, leaving conventional musical education largely obsolete and enabling contributions by musically untrained scientists. As this essay has illustrated, music did not lose its relevance entirely. Alluring gramophone records of bird sound attracted amateurs, who then contributed to data collection in ornithology. Musical metaphors continued to be used in both the didactics and the analysis of sound. Moreover, some bioacousticians added musical notation to sound spectrograms; others even hired a drum-master to distinguish and identify the rhythmic structures in the inter-click intervals of sperm whales (Harris 2012). And in Harris and Flynn's study, medical students who had received musical training earlier in life thought this helped them to discriminate between bodily sounds and find words for them. As one student explained, talking about lung sounds: "If you play music you listen to a lot of music and you become really particular about … you hear the phrasing. You can hear the rests, you can hear the pauses. You become more skilled in listening to the nuances of sound" (S7, cited in Harris and Flynn 2017: 4). A musically inspired skills training—focusing on the discrimination of rhythm, cadence, pitch, and timbre—enhanced nursing students' ability to notice and interpret the sounds of heart, lungs, and gut, such as the increase of bowel sounds "with diarrhea, gastroenteritis, or early obstruction" (Pellico et al. 2012: 236). In all, however, musical skills have tended to lose their analytic value in the sciences over the late nineteenth, twentieth and early twenty-first century.

How, then, should we understand scientists' currently growing interest in musical sonification and the evocation of the sublime that it promises to deliver? Alexandra Supper has cited the rise of interdisciplinarity in the sciences, noting the explanation of this recent prominence by sociologist of science Andrew Barry et al. (2008). They claim that interdisciplinarity—which in their definition includes the integration of distinct fields of academic scholarship and the integration of science with non-science, such as the arts—draws on three logics: the logic of innovation, the logic of accountability, and the logic of ontology. The idea that the sciences should contribute to technological innovation and economic growth, or the logic of innovation, is now acclaimed within and beyond academia, and has propelled all kinds of research that combines insights and methods from different disciplines to come up with novel medicines or other products for the market. In contrast, the logic of ontology is geared to "effecting ontological change in both the object(s) of research, and the relations between research subjects and objects" (Born and Barry 2010: 105). Reflection on the reality claims behind particular methods, or on the shifting boundaries between humans and technology, pushes academics toward interdisciplinarity, which may then further erode conventional ontologies. The logic of accountability is anchored in the ever greater need to account for the large amounts of public spending on science, and to respond to the declining authority of and a growing public "unease" with scientific institutions. One way for scientists to cope with these pressures is by "enlisting artists" to reach out for audiences who "might develop not only cognitive, but interactive and affective involvements with science" (Born and Barry 2010: 108–109). Interdisciplinary projects of this kind are intended to legitimize investment in and by scientists.

The increasing demand for science's accountability expresses a major shift in the relationships between science and society—a shift that is also an important factor for the rise of sonification projects in which scientists and artists collaborate. But scientists have not been the only ones seeking to enhance the legitimacy of their work through strategic alliances. Artists have turned to the sciences with the aim of making their own work more acceptable or visible. This is reflected in the hope expressed by artist John Dunn, mentioned in passing in the previous chapter, of adding "a sense of 'deep structure'" to music when he worked with biologist Mary Anne Clark on turning protein sequences into a work of sound art (Supper 2014: 43). This two-way strategic referencing phenomenon is known as "legitimacy exchange," a term coined by STS scholar Geof Bowker (1993: 116). The notion indicates how scientists at the margins of particular fields refer to colleagues in other domains of science in order to garner support for their claims. Some sonification projects have even wider ambitions, though. The composition *Bonner Durchmusterung* by Marcus Schmickler, commissioned for the International Year of Astronomy in 2009, was intended to educate the public about astronomical phenomena such as eruptions of the sun. But the program notes, co-authored by the scientists and artists involved, say that the work also proposes to foster reflection about "the relation between data and the reality of the observed objects" (cited by Supper 2014: 44)—reminding us of the logic of ontology.

It remains to be seen whether this particular instance of sonification will help to consolidate the trustworthiness of science in the public eye, but many sonifications seem well suited to creating a new public engagement *with* science, beyond a more conventional public understanding *of* science (Supper 2014: 50). The need for such engagement is particularly acute now that governmental budget cuts increasingly threaten science, the steeply rising expenditure of which is no longer considered self-evident. This situation calls for new and dynamic ways of emphasizing the value of science, and sensory immersion in technologically mediated and artistic inspired sonifications is one such way. It is another, and revealing, ensemble of sonic skills in the sciences. As one would expect, the need for accountability prompted various ways of monitory listening in labs and hospitals—but it also gave a boost to sonifications for wider audiences that tapped into exploratory-synthetic listening through technology-enhanced immersion.

Epilogue

We experienced an accountability driven and aesthetically informed ensemble of sonic skills ourselves in the aftermath of the Sonic Skills project. When most of the research had been completed, in 2015 we organized a Sonic Science Festival with lectures, demonstrations, concerts, a children's workshop, and an exhibition. The festival aimed to present our own research on sound and listening in science, medicine, and engineering, but also to spark the interest of children, young adults, and others in science more generally, through sound and music. For one of the concerts, we collaborated with the Maastricht sound art venue Intro in Situ, and commissioned the young guitar player and composer Aart Strootman to create a piece for his "minimal chamber metal" band Temko. He prepared for his composition by reading the entire *The Oxford Handbook of Sound Studies* (Pinch and Bijsterveld 2012) and Alexandra Supper's 2014 article on sublime frequencies.

This inspired Strootman to start experimenting with sonification in composition. The first thing he did was to send an email to the US National Aeronautics and Space Administration (NASA). He requested—successfully, to his own surprise—the last set of data that NASA ever received from space probe Pioneer 10. This spacecraft was launched in 1972 for a mission that would last for over thirty years, until the last moment of contact between NASA and Pioneer 10 on January 22, 2003. At that moment, the distance between Pioneer 10 and the earth was 8 bn kilometers. Strootman's second step was to transform the dataset into the composition *Darkness Rises*. As well as having the data mold the structure of his musical materials, he expressed time's passing using visually displayed information and radio broadcasts on important events in the political, cultural, scientific, and technological history of humankind. "Sonic Skills paved new paths in my personal composition practice," Strootman concluded. "A scientific approach towards music, in the shape of sonification, has become a ubiquitous component in my writing. 'Darkness Rises' is the first but important step into this territory."[2]

The composition immediately attracted the interest of programmers at venues for classical, avantgarde, jazz, and pop music, and was played on eleven occasions. Among them were a concert at the Amsterdam Concertgebouw, a Rotterdam event that showcased *Darkness Rises* for the Classical: NEXT forum, and a performance in Eindhoven that also featured public intellectual Bas Heijne speaking on the future of science. This last event included the release of Temko's recording of *Darkness Rises*. But there was more to come: the piece lived on in a version for symphony orchestra. The Philharmonie Zuidnederland programmed *Darkness Rises* in its Spicy Classics series along with classical music concerts "in a modern jacket," a "cross-over" to be enjoyed "with a beer in one's hand," reaching out for new audiences. In its advertising for the concert, the orchestra stressed that the musically rendered dataset "marks the last thing we ... heard from Pioneer 10 .... It is now travelling through space's gigantic void for eternity."[3]

This is a perfect example of rhetorically promising the auditory sublime. What started as our way of bringing science to a wider audience, ended up bringing classical music to new publics with a musical sonification of scientific data. It was the result of a legitimacy exchange *par excellence*, as music legitimated science's existence while science underscored the vitality of classical music in new formats. The composition was the fruit of an ensemble of sonic skills that we expect to stay around as long as accountability remains important in both the sciences and the arts.

We would have been very unlikely to trace ensembles like this if we had limited our research to historical research alone. Our ethnographies of labs, factories, hospitals, and the sonification community, our reenactments of sound and listening in the sciences, and the long-term relationships we were able to forge with scientists, engineers, doctors, and artists alike were also highly informative. By pairing a sounding history of science with a "sounded anthropology" (Samuels et al. 2010; Bijsterveld 2016), we learned as much about sound in science as about the dynamics between science and technology, science and professions, and science and society. If the history of sound can "disclose previously unknown historical connections," as historian Daniel Morat recapped the ambitions of his peer Mark M. Smith (Morat 2014: 2–3), then our case studies of the sciences through the lens of sound foregrounded the continued significance of trust, and the renewed significance of timing and accountability, in the sciences. They helped us to identify the vitality of the embodied, sound-informed synchronization work required to deal with the tensions between organizational time and expensive instrumental time in today's large-scale laboratories. They enabled us to claim that some sonic skills long ago escaped the trend toward standardization and visualization, or "popped up" again in situations where listening was embedded in legally sanctioned or organizationally invested structures of trust. And they showed us that attempts to invoke the sublime through sonification respond to society's increasing demands for both scientists and artists to account for what they do—by engaging with, rather than conventionally educating, a wide audience. It was a pleasure to listen to them.

NOTES

1. Personal communication André Kuipers to Karin Bijsterveld, November 29, 2017, courtesy André Kuipers.
2. Personal communication Aart Strootman to Alexandra Supper, January 22, 2017.
3. http://www.philharmoniezuidnederland.nl/concerts/spicy-classics-music-space/ (last accessed January 31, 2017).

REFERENCES

Barry, A., Born, G., & Weszkalnys, G. (2008). Logics of Interdisciplinarity. *Economy and Society, 37*(1), 20–49.
Bijsterveld, K. (2016). Ethnography and Archival Research in Studying Cultures of Sound. In H. Schulze & J. Papenburg (Eds.), *Sound as Popular Culture: A Research Companion* (pp. 99–109). Cambridge: MIT Press.
Born, G., & Barry, A. (2010). Art-Science: From Public Understanding to Public Experiment. *Journal of Cultural Economy, 3*(1), 103–119.
Bowker, G. (1993). How to Be Universal: Some Cybernetic Strategies, 1943–1970. *Social Studies of Science, 23*(1), 107–127.
Bruyninckx, J. (2017). Synchronicity: Time, Technicians, Instruments, and Invisible Repair. *Technology & Human Values, 42*(5), 822–847.
Bruyninckx, J., & Supper, A. (2016). Sonic Skills in Cultural Contexts: Theories, Practices and Materialities of Listening. *Sound Studies: An Interdisciplinary Journal, 2*(1), 1–5.
Harris, A. (2016). Listening-Touch, Affect and the Crafting of Medical Bodies Through Percussion. *Body & Society, 22*(1), 31–61.
Harris, Y. (2012). Understanding Underwater: The Art and Science of Interpreting Whale Sounds. *Interference: A Journal of Audio Culture*. Available at http://www.interferencejournal.org/understanding-underwater/. Last accessed August 18, 2017.
Harris, A. (2015). Sounding Disease: Guest Blog at Sociology of Diagnosis Website. Available at https://www.facebook.com/SociologyOfDiagnosis/posts/799049830181091. Last accessed August 18, 2017.
Harris, A., & Flynn, E. (2017). Medical Education of Attention: A Qualitative Study of Learning to Listen to Sound. *Medical Teacher, 39*(1), 79–84.
Hui, A. (2013). *The Psychophysical Ear: Musical Experiments, Experimental Sounds, 1840–1910*. Cambridge: MIT Press.
Morat, D. (2014). Introduction. In D. Morat (Ed.), *Sounds of Modern History: Auditory Cultures in 19th- and 20th-Century Europe* (pp. 1–9). New York, NY: Berghahn.
Pellico, L., Duffy, T., Fennie, K., & Swan, K. (2012). Looking Is Not Seeing and Listening Is Not Hearing: Effect of an Intervention to Enhance Auditory

Skills of Graduate-Entry Nursing Students. *Nursing Education Perspectives*, *33*(4), 234–239.

Pinch, T., & Bijsterveld, K. (Eds.). (2012). *The Oxford Handbook of Sound Studies*. Oxford: Oxford University Press.

Rice, T. (2016). Sounds Inside: Prison, Prisoners and Acoustical Agency. *Sound Studies*, *2*(1), 6–20.

Roosth, S. (2009). Screaming Yeast: Sonocytology, Cytoplasmic Milieus, and Cellular Subjectivities. *Critical Inquiry*, *35*(2), 332–350.

Samuels, D. W., Meintjes, L., Ochoa, A. M., & Porcello, T. (2010). Soundscapes: Toward a Sounded Anthropology. *Annual Review of Anthropology*, *39*, 329–345.

Supper, A. (2014). Sublime Frequencies: The Construction of Sublime Listening Experiences in the Sonification of Scientific Data. *Social Studies of Science*, *44*(1), 34–58.

Supper, A. (2015). Data Karaoke: Sensory and Bodily Skills in Conference Presentations. *Science as Culture*, *24*(4), 436–457.

Supper, A. (2016). Lobbying for the Ear, Listening with the Whole Body: The (Anti-)Visual Culture of Sonification. *Sound Studies: An Interdisciplinary Journal*, *2*(1), 69–80.

REFERENCES

Abbott, A. (1988). *The System of Professions: An Essay on the Division of Expert Labour*. Chicago: University of Chicago Press.

Abraham, O., & Von Hornbostel, Erich M. (1909). Vorschläge für die Transkription exotischer Melodien. *Sammelbände der Internationalen Musikgesellschaft, 11*(1), 1–25.

Adorno, T. W. (1977/1962). Typen musikalischen Verhaltens. In *Einleitung in die Musiksoziologie* (pp. 14–34). Frankfurt am Main: Suhrkamp Verlag.

Alač, M. (2014). Digital Scientific Visuals as Fields for Interaction. In C. Coopmans, J. Vertesi, M. Lynch, & S. Woolgar (Eds.), *Representation in Scientific Practice Revisited* (pp. 61–87). Cambridge: MIT Press.

Alberts, G. (2000). Computergeluiden. In G. Alberts & R. van Dael (Eds.), *Informatica & Samenleving* (pp. 7–9). Nijmegen: Katholieke Universiteit Nijmegen.

Alberts, G. (2003). Een halve eeuw computers in Nederland. *Nieuwe Wiskrant, 22*(3), 17–23.

Amsterdamska, O. (2005). Demarcating Epidemiology. *Science, Technology, and Human Values, 30*(1), 17–51.

Baier, G., Hermann, T., & Stephani, U. (2007). Multi-channel Sonification of Human EEG. In *Proceedings of the 13th International Conference on Auditory Display, Montreal, Canada, June 26–29* (pp. 491–496).

Bailey, P. (1996). Breaking the Sound Barrier: A Historian Listens to Noise. *Body & Society, 2*(2), 49–66.

Barany, M. J., & MacKenzie, D. (2014). Chalk: Materials and Concepts in Mathematics Research. In C. Coopmans, J. Vertesi, M. Lynch, & S. Woolgar (Eds.), *Representation in Scientific Practice Revisited* (pp. 107–129). Cambridge: MIT Press.

© The Editor(s) (if applicable) and The Author(s) 2019
K. Bijsterveld, *Sonic Skills*,
https://doi.org/10.1057/978-1-137-59829-5

Barry, A., Born, G., & Weszkalnys, G. (2008). Logics of interdisciplinarity. *Economy and Society, 37*(1), 20–49.

Bijsterveld, K. (2007). *Weg van geluid: Hoe de auto een plaats werd om tot rust te komen.* Maastricht: Universiteit Maastricht.

Bijsterveld, K. (2008). *Mechanical Sound: Technology, Culture and Public Problems of Noise in the Twentieth Century.* Cambridge: MIT Press.

Bijsterveld, K. (2012). Listening to Machines: Industrial Noise, Hearing Loss and the Cultural Meaning of Sound. In T. S. S. Reader (Ed.), *Jonathan Sterne* (pp. 152–167). New York: Routledge.

Bijsterveld, K. (2016). Ethnography and Archival Research in Studying Cultures of Sound. In H. Schulze & J. Papenburg (Eds.), *Sound as Popular Culture: A Research Companion* (pp. 99–109). Cambridge: MIT Press.

Bijsterveld, K., & Krebs, S. (2013). Listening to the Sounding Objects of the Past: The Case of the Car. In K. Franinović & S. Serafin (Eds.), *Sonic Interaction Design* (pp. 3–38). Cambridge: MIT Press.

Bijsterveld, K., Cleophas, E., Krebs, S., & Mom, G. (2014). *Sound and Safe: A History of Listening Behind the Wheel.* Oxford: Oxford University Press.

Borg, K. (2007). *Auto Mechanics: Technology and Expertise in Twentieth-Century America.* Baltimore, MD: Johns Hopkins University Press.

Born, G., & Barry, A. (2010). Art-Science: From Public Understanding to Public Experiment. *Journal of Cultural Economy, 3*(1), 103–119.

Bourdieu, P. (1984). *Distinction: A Social Critique of the Judgement of Taste.* Cambridge, MA: Harvard University Press.

Bowker, G. (1993). How to Be Universal: Some Cybernetic Strategies, 1943–70. *Social Studies of Science, 23*(1), 107–127.

Brady, E. (1999). *A Spiral Way: How the Phonograph Changed Ethnography.* Jackson: University Press of Mississippi.

Brăiloiu, C. (1970/1931). Outline of a Method of Musical Folklore. *Ethnomusicology, 14*(3), 389–417.

Bregman, A. (1994). *Auditory Scene Analysis: The Perceptual Organization of Sound.* Cambridge: MIT Press.

Broeders, A. P. A. (2002). Het herkennen van stemmen. In P. J. van Koppen, et al. (Eds.), *Het recht van binnen: psychologie van het recht* (pp. 573–596). Deventer: Kluwer.

Bronfman, A. (2016). Biography of a Sonic Archive. *Hispanic American Historical Review, 96*(2), 225–231.

Bruton, E., & Gooday, G. (2016a). Listening in Combat: Surveillance Technologies Beyond the Visual in the First World War. *History and Technology, 32*(3), 213–226.

Bruton, E., & Gooday, G. (2016b). Listening in the Dark: Audio Surveillance, Communication Technologies, and the Submarine Threat During the First World War. *History and Technology, 32*(3), 245–268.

Bruyninckx, J. (2012). Sound Sterile: Making Scientific Field Recordings in Ornithology. In T. Pinch & K. Bijsterveld (Eds.), *The Oxford Handbook of Sound Studies* (pp. 127–150). Oxford: Oxford University Press.

Bruyninckx, J. (2013). Sound Science: Recording and Listening in the Biology of Bird Song, 1880–1980 (Ph.D. thesis, Maastricht University).

Bruyninckx, J. (2014). Silent City: Listening to Birds in Urban Nature. In M. Gandy & B. Nilsen (Eds.), *The Acoustic City* (pp. 42–48). Berlin: Jovis.

Bruyninckx, J. (2015). Trading Twitter: Amateur Recorders and Economies of Scientific Exchange at the Cornell Library of Natural Sounds. *Social Studies of Science, 45*(3), 344–370.

Bruyninckx, J. (2017). Synchronicity: Time, Technicians, Instruments, and Invisible Repair. *Science, Technology & Human Values, 42*(5), 822–847.

Bruyninckx, J. (2018a). *Listening in the Field: Recording and the Science of Birdsong.* Cambridge: MIT Press.

Bruyninckx, J. (2018b). Instrument Trust and Somatic Vigilance in Experimental Physics. *Science as Culture.*

Bruyninckx, J., & Supper, A. (2016). Sonic Skills in Cultural Contexts: Theories, Practices and Materialities of Listening. *Sound Studies: An Interdisciplinary Journal, 2*(1), 1–5.

Bull, M. (2006). Auditory. In C. A. Jones (Ed.), *Sensorium: Embodied Experience, Technology, and Contemporary Art* (pp. 112–114). Cambridge: MIT Press.

Bull, M., & Back, L. (Eds.). (2003). *The Auditory Culture Reader.* Oxford: Berg.

Burri, R. V. (2008). Doing Distinctions: Boundary Work and Symbolic Capital in Radiology. *Social Studies of Science, 38*(1), 35–62.

Burri, R. V., Schubert, C., & Strübing, J. (2011). Introduction: The Five Senses of Science. *Science, Technology & Innovation Studies, 7*(1), 3–7.

Busoni, F. (1962/1907). Sketch of a New Esthetic of Music. In *Three Classics in the Aesthetic of Music* (pp. 75–102). New York: Dover Publications.

Caravaglios, C. (1935). The Collection and Transcription of Folk-Dances. *Journal of the English Folk Dance and Song Society, 2*, International Festival Number, 127–135.

Chion, M. (2005/1990). *Audio-Vision: Sound on Screen.* New York: Columbia University Press.

Classen, C. (1997). Foundations for an Anthropology of the Senses. *International Social Science Journal, 49*(153), 401–412.

Cleophas, E., & Bijsterveld, K. (2012). Selling Sound: Testing, Designing and Marketing Sound in the European Car Industry. In T. Pinch & K. Bijsterveld (Eds.), *The Oxford Handbook of Sound Studies* (pp. 102–124). Oxford: Oxford University Press.

Coghlan, A. (2014, June 12). Massive "Ocean" Discovered Towards Earth's Core. *New Scientist.* Available at http://www.newscientist.com/article/

dn25723-massive-ocean-discovered-towards-earths-core.html?cmpid=RSS|N-SNS|2012-GLOBAL|online-news#.VK_BGyx0z9L. Last accessed March 13, 2015.

Collins, H. M. (1985). *Changing Order: Replication and Induction in Scientific Practice*. London: Sage.

Collins, H. M. (2001). Tacit Knowledge, Trust, and the Q of Sapphire. *Social Studies of Science, 31*(1), 71–86.

Collins, H. M. (2013). Building an Antenna for Tacit Knowledge. In L. Soler, S. D. Zwart, & R. Catinaud (Eds.), Tacit and Explicit Knowledge: Harry Collins's Framework. *Philosophia Scientiae, 17*(3), 25–39.

Corbin, A. (1999). *Village Bells: Sound and Meaning in the Nineteenth-Century French Countryside*. London: Macmillan.

Cowell, H. (1932). Henry Cowell schrijft ons. *Maandblad voor Hedendaagsche Muziek, 1*(12), 90–91.

Daston, L., & Galison, P. (1992). The Image of Objectivity. *Representations, 10*(40), 81–128.

Daston, L., & Galison, P. (2007). *Objectivity*. New York: Zone Books.

Dayé, C., & de Campo, A. (2006). Sounds Sequential: Sonification in the Social Sciences. *Interdisciplinary Science Reviews, 31*(4), 349–364.

Dell'Antonio, A. (Ed.). (2004). *Beyond Structural Listening? Postmodern Modes of Hearing*. Berkeley: University of California Press.

Dombois, F. (2001). Using Sonification in Planetary Seismology. In *Proceedings of the 7th International Conference on Auditory Display, Espoo, Finland, July 29–August 1* (pp. 227–230).

Douglas, M. (1937). Manx Folk Dances: Their Notation and Revival. *Journal of the English Folk Dance and Song Society, 3*(2), 110–116.

Douglas, S. J. (1999). *Listening in: Radio and the American Imagination, from Amos 'n' Andy and Edward R. Murrow to Wolfman Jack and Howard Stern*. New York: Times Books.

Duffin, J. (1998). *To See with a Better Eye: A Life of R.T.H. Laennec*. Princeton, NJ: Princeton University Press.

Edworthy, J., & Hards, R. (1999). Learning Auditory Warnings: The Effects of Sound Type, Verbal Labelling and Imagery on the Identification of Alarm Sounds. *International Journal of Industrial Ergonomics, 24*(6), 603–618.

Encke, J. (2006). *Augenblicke der Gefahr: Der Krieg und die Sinne*. München: Wilhelm Fink Verlag.

Endsley, M. R., & Jones, D. G. (2012). *Designing for Situation Awareness: An Approach to User-Centered Design* (2nd ed.). Boca Raton, CA: CRC Press.

Erlmann, V. (2010). *Reason and Resonance: A History of Modern Aurality*. New York, NY: Zone Books.

Fehr, J. (2000). "Visible Speech" and Linguistic Insight. In H. Nowotny & M. Weiss (Eds.), *Shifting Boundaries of the Real: Making the Invisible Visible* (pp. 31–47). Zürich: VDF Hochschulverlag AG an der ETH Zürich.

Feld, S. (2003). A Rainforest Acoustemology. In T. A. C. Reader (Ed.), *Michael Bull & Les Back* (pp. 223–239). Oxford: Berg.

Feld, S., & Brenneis, D. (2004). Doing Anthropology in Sound. *American Ethnologist, 31*(4), 461–474.

Ferguson, E. S. (1992). *Engineering and the Mind's Eye.* Cambridge: MIT Press.

Flowers, J. (2005). Thirteen Years of Reflection on Auditory Graphing: Promises, Pitfalls, and Potential New Directions. In *Proceedings of the 11th International Conference on Auditory Display, Limerick, Ireland, July 6–9* (pp. 405–409).

Gallop, R. (1935). Systematization of Motives in the Ceremonial Dance. *Journal of the English Folk Dance and Song Society, 2,* International Festival Number, 79–83.

Garfinkel, H. (1967). *Studies in Ethnomethodology.* Englewood Cliffs: Prentice Hall.

Gaver, W. W. (1989). The SonicFinder: An Interface That Uses Auditory Icons. *Human-Computer Interaction, 4*(1), 67–94.

Gautier, A. M. O. (2014). *Aurality: Listening & Knowledge in Nineteenth-Century Colombia.* Durham, NC: Duke University Press.

Gerard, P. (2002). *Secret Soldiers: The Story of World War II's Heroic Army of Deception.* New York, NY: Dutton.

Gibling, S. P. (1917). Types of Musical Listening. *Musical Quarterly, 3*(3), 385–389.

Gieryn, T. F. (1995). Boundaries of Science. In S. Jasanoff, G. E. Markle, J. C. Petersen, & T. J. Pinch (Eds.), *Handbook of Science and Technology Studies* (pp. 393–443). Thousand Oaks, CA: Sage.

Gitelman, L. (1999). *Scripts, Grooves, and Writing Machines: Representing Technology in the Edison Era.* Stanford, CA: Stanford University Press.

Glowacki, O., Deane, G. B., Moskalik, M., Blondel, P., Tegowski, J., & Blaszczyk, M. (2015). Underwater Acoustic Signatures of Glacier Calving. *Geophysical Research Letters, 42*(3), 804–812.

Goodman, D. (2010). Distracted Listening: On Not Making Sound Choices in the 1930s. In D. Suisman & S. Strasser (Eds.), *Sound in the Age of Mechanical Reproduction* (pp. 15–46). Philadelphia: University of Pennsylvania Press.

Goodwin, S. (2010). *Sonic Warfare: Sound, Affect, and the Ecology of Fear.* Cambridge: MIT Press.

Grond, F., & Hermann, T. (2014). Interactive Sonification for Data Exploration: How Listening Modes and Display Purposes Define Design Guidelines. *Organised Sound, 19*(1), 41–51.

Hackmann, W. (1984). *Seek and Strike: Sonar, Anti-submarine Warfare and the Royal Navy, 1914–54*. London: Her Majesty's Stationery Office.

Halffman, W. (2003). *Boundaries of Regulatory Science: Eco/Toxicology and Aquatic Hazards of Chemicals in the US, England, and the Netherlands* (Ph.D. thesis, University of Amsterdam).

Harris, Y. (2012). Understanding Underwater: The Art and Science of Interpreting Whale Sounds. *Interference: A Journal of Audio Culture*. Available at http://www.interferencejournal.org/understanding-underwater/. Last accessed August 18, 2017.

Harris, A. (2015a). Eliciting Sound Memories. *The Public Historian, 37*(4), 14–31.

Harris, A. (2015b). Autophony: Listening to Your Eyes Move. *Somatosphere: Science, Medicine and Anthropology*. Available at http://somatosphere. net/2015/06/autophony-listening-to-your-eyes-move.html. Last accessed August 18, 2017.

Harris, A. (2015c). *Sounding Disease: Guest Blog at Sociology of Diagnosis Website*. Available at https://www.facebook.com/SociologyOfDiagnosis/posts/799049830181091. Last accessed August 18, 2017.

Harris, A. (2016). Listening-Touch, Affect and the Crafting of Medical Bodies Through Percussion. *Body & Society, 22*(1), 31–61.

Harris, A. & Van Drie, M. (2015). Sharing Sound: Teaching, Learning and Researching Sonic Skills. *Sound Studies: An Interdisciplinary Journal, 1*(1), 98–117.

Harris, A., & Flynn, E. (2017). Medical Education of Attention: A Qualitative Study of Learning to Listen to Sound. *Medical Teacher, 39*(1), 79–84.

Heering, P. (2008). The Enligthened Microscope: Re-enactment and Analysis of Projections with Eighteenth-Century Solar Microscopes. *British Journal for the History of Science, 41*(3), 345–367.

Helmreich, S. (2012). Underwater Music: Tuning Composition to the Sounds of Science. In T. Pinch & K. Bijsterveld (Eds.), *The Oxford Handbook of Sound Studies* (pp. 151–175). Oxford: Oxford University Press.

Hermann, T. (2002). *Sonification for Exploratory Data Analysis* (Ph.D. thesis, Bielefeld University).

Hermann, T. (2011). Model-Based Sonification. In T. Hermann, A. Hunt, & J. G. Neuhoff (Eds.), *The Sonification Handbook* (pp. 399–427). Berlin: Logos Verlag.

Hermann, T., & Hunt, A. (2005). An Introduction to Interactive Sonification. *IEEE Multimedia, 12*(2), 20–24.

Hermann, T., & Hunt, A. (2011). Interactive Sonification. In T. Hermann, A. Hunt, & J. G. Neuhoff (Eds.), *The Sonification Handbook* (pp. 273–298). Berlin: Logos Verlag.

Hermann, T., Hunt, A., & Neuhoff, J. G. (2011). Introduction. In T. Sonification (Ed.), *Handbook* (pp. 1–6). Berlin: Logos Verlag.

Hilmes, M. (2005). Is There a Field Called Sound Culture Studies? And Does It Matter? *American Quarterly, 57*(1), 249–259.

Hochman, B. (2010). Hearing Lost, Hearing Found: George Washington Cable and the Phono-Ethnographic Ear. *American Literature, 82*(3), 519–551.

Hoffmann, C. (1994). Wissenschaft und Militär: Das Berliner Psychologische Institut und der I. Weltkrieg. *Psychologie und Geschichte, 5*(3–4), 261–285.

Hui, A. (2013). *The Psychophysical Ear: Musical Experiments, Experimental Sounds, 1840–1910*. Cambridge: MIT Press.

Hui, A., Kursell, J., & Jackson, M. (Eds.). (2013). Music, Sound and the Laboratory from 1750–1980. *Osiris, 28*(1), 1–11.

Ingold, T. (2000). *The Perception of the Environment: Essays on Livelihood, Dwelling and Skill*. London: Routledge.

Ingold, T. (2007). *Lines: A Brief History*. Milton Park: Routledge.

Ingold, T. (2011a). Worlds of Sense and Sensing the World: A Response to Sarah Pink and David Howes. *Social Anthropology, 19*(3), 313–317.

Ingold, T. (2011b). Reply to David Howes. *Social Anthropology, 19*(3), 323–327.

Jannin, J. (1852). *L'Art d'élever et de multiplier les serins canaris et hollandais*. Paris: Chez Tissot.

Johns, A. (1998). *The Nature of the Book: Print and Knowledge in the Making*. Chicago, IL: University of Chicago Press.

Judkins, P. (2016). Sound and Fury: Sound and Vision in Early U.K. Air Defence. *History and Technology, 32*(3), 227–244.

Kane, B. (2015). Sound Studies Without Auditory Culture: A Critique of the Ontological Turn. *Sound Studies: An Interdisciplinary Journal, 1*(1), 2–21.

Karwoski, T. F., & Odbert, H. S. (1938). Color-Music. *Psychological Monographs, 50, 2*.

Kenworthy Schofield, R. (1928). Morris Dances from Field Town. *Journal of the English Folk Dance Society, 2*, 22–28.

Klein, J. T. (1996). *Crossing Boundaries: Knowledge, Disciplinarities, and Interdisciplinarities*. Charlottesville: University Press of Virginia.

Korpáš, J., Sadloňová, J., & Vrabec, M. (1996). Analysis of the Cough Sound: An Overview. *Pulmonary Pharmacology, 9*(5–6), 261–268.

Kramer, G. (Ed.). (1999). *Sonification Report: Status of the Field and Research Agenda, International Community for Auditory Display*. Available at http://icad.org/websiteV2.0/References/nsf.html. Last accessed August 18, 2017.

Krebs, S. (2012a). Automobilgeräusche als Information: Über das geschulte Ohr des Kfz-Mechanikers. In A. Schoon & A. Volmar (Eds.), *Das geschulte Ohr: Eine Kulturgeschichte der Sonifikation* (pp. 95–110). Bielefeld: Transcript.

Krebs, S. (2012b). "Notschrei eines Automobilisten" oder die Herausbildung des Kfz-Handwerks in Deutschland. *Technikgeschichte, 79*(3), 185–206.

Krebs, S. (2012c). "Sobbing, Whining, Rumbling": Listening to Automobiles as Social Practice. In T. Pinch & K. Bijsterveld (Eds.), *The Oxford Handbook of Sound Studies* (pp. 79–101). Oxford: Oxford University Press.

Krebs, S. (2013). Von Motorkonzerten und aristokratischer Stille: Die Einführung der geschlossenen Automobilkarosserie in Frankreich und Deutschland, 1919–1939. In R.-J. Gleitsmann & J. Wittmann (Eds.), *Innovationskulturen um das Automobil: Von gestern bis morgen* (pp. 77–99). Königswinter: Heel.

Krebs, S. (2014a). "Dial Gauge versus Sense 1–0": German Car Mechanics and the Introduction of New Diagnostic Equipment, 1950–1980. *Technology and Culture, 55*(2), 354–389.

Krebs, S. (2014b). Diagnose nach Gehör? Die Aushandlung neuer Wissensformen in der Kfz-Diagnose (1950–1980), *Ferrum: Wissensformen der Technik, 86,* 79–88.

Krebs, S. (2015). Einleitung: Zur Sinnlichkeit der Technik(geschichte). Ist es Zeit für einen »sensorial turn«? *Technikgeschichte, 82*(1), 3–10.

Krebs, S. (2017). Memories of a Dying Industry: Sense and Identity in a British Paper Mill. *The Senses & Society, 12*(1), 35–52.

Krebs, S., & Van Drie, M. (2014). The Art of Stethoscope Use: Diagnostic Listening Practices of Medical Physicians and "Auto Doctors", *ICON: Journal of the International Committee for the History of Technology 20*(2), 92–114.

Kursell, J. (Ed.). (2008). *Sounds of Science-Schall im Labor (1800–1930).* Berlin: Max Planck Institut für Wissenschafsgeschichte.

Labelle, B. (2011). *Acoustic Territories: Sound Culture and Everyday Life.* New York, NY: Continuum.

Lacey, K. (2013). *Listening Publics: The Politics and Experience of Listening in the Media Age.* Cambridge: Polity Press.

Lachmund, J. (1994). *Der abgehorchte Körper: Zur historischen Soziologie der medizinischen Untersuchung.* Opladen: Westdeutscher Verlag.

Lachmund, J. (1999). Making Sense of Sound: Auscultation and Lung Sound Codification in Nineteenth-Century French and German Medicine. *Science, Technology, & Human Values, 24*(4), 419–450.

Landi, E., Alexander, R. L., Gruesbeck, J. R., Gilbert, J. A., Lepri, S. T., Manchester, W. B., ..., Zurbuchen, T. H. (2012). Carbon Ionization Stages as a Diagnostic of the Solar Wind. *The Astrophysical Journal, 744* (2), 100.

Latour, B. (1986). Visualisation and Cognition: Thinking with Eyes and Hands. *Knowledge and Society: Studies in the Sociology of Culture Past and Present, 6,* 1–40.

Latour, B. (1987). *Science in Action: How to Follow Scientists and Engineers Through Society*. Milton Keynes: Open University Press.

Lethen, H. (2000). "Knall an sich": Das Ohr als Einbruchstelle des Traumas. In I. Mülder-Bach (Ed.), *Modernität und Trauma: Beiträge zum Zeitenbruch des Ersten Welkrieges* (pp. 192–210). Wien: Universitätsverlag.

Lorimer, J. (2008). Counting Corncrakes: The Affective Science of the UK Corncrake Census. *Social Studies of Science, 38*(3), 377–405.

Lynch, M. (1990). The Externalized Retina: Selection and Mathematization in the Visual Documentation of Objects in the Life Sciences. In M. Lynch & S. Woolgar (Eds.), *Representation in Scientific Practice* (pp. 153–186). Cambridge: MIT Press.

Lynch, M. (2013). At the Margins of Tacit Knowledge. In L. Soler, S. D. Zwart, & R. Catinaud (Eds.), *Tacit and Explicit Knowledge: Harry Collins's Framework. Philosophia Scientiae, 17*(3), 55–73.

Mahrenholz, J.-K. (2008). Etnografische geluidsopnames in Duitse krijgsgevangenenkampen tijdens de Eerste Wereldoorlog. In D. Dendooven & P. Chielen (Red.), *Vijf continenten in Vlaanderen* (pp. 161–165). Tielt: Lannoo.

Maisonneuve, S. (2001). Between History and Commodity: The Production of a Musical Patrimony Through the Record in the 1920s–1930s. *Poetics, 29*(2), 89–108.

Marsden, W. J. M. (1927). Some Observations on Bird Music. *Music & Letters, 8*(3), 339–344.

Martin, M., & Fangerau, H. (2011). Töne sehen? Zur Visualisierung akustischer Phänomene in der Herzdiagnostik. *NTM Zeitschrift fur Geschichte der Wissenschaften, Technik und Medizin, 19*(3), 299–327.

Maslen, S. (2015). Researching the Senses as Knowledge: A Case Study of Learning to Hear Medically. *The Senses & Society, 10*(1), 52–70.

Mayer, K. (2011). Scientific Images? How Touching! *Science, Technology & Innovation Studies, 7*(1), 29–45.

Mills, M. (2010). Deaf Jam: From Inscription to Reproduction to Information. *Social Text, 28*(1) (102), 35–58.

Mody, C. C. M. (2005). The Sounds of Science: Listening to Laboratory Practice. *Science, Technology & Human Values, 30*(2), 175–198.

Mody, C. C. M. (2012). Conversions: Sound and Sight, Military and Civilian. In T. Pinch & K. Bijsterveld (Eds.), *The Oxford Handbook of Sound Studies* (pp. 224–248). Oxford: Oxford University Press.

Mody, C. C. M. (2014). Essential Tensions and Representational Strategies. In C. Coopmans, J. Vertesi, M. Lynch, & S. Woolgar (Eds.), *Representation in Scientific Practice Revisited* (pp. 223–248). Cambridge: MIT Press.

Morat, D. (2014). Introduction. In D. Morat (Ed.), *Sounds of Modern History: Auditory Cultures in 19th- and 20th-Century Europe* (pp. 1–9). New York, NY: Berghahn.

Mundy, R. (2009). Birdsong and the Image of Evolution. *Society and Animals*, *17*(3), 206–223.

Mundy, R. (2010). *Nature's Music: Birds, Beasts, and Evolutionary Listening in the Twentieth Century* (Ph.D. thesis, New York University).

Mundy, R. (2018). *Animal Musicalities: Birds, Beasts, and Evolutionary Listening*. Middletown, CO: Wesleyan University Press.

Myers, N. (2008). Molecular Embodiments and the Body-Work of Modeling in Protein Chrystallography. *Social Studies of Science*, *38*(2), 163–199.

Orr, J. E. (1996). *Talking About Machines: An Ethnography of a Modern Job*. Ithaca and London: Cornell University Press.

Osgood, C. E., Suci, G. J., & Tannenbaum, P. H. (1957). *The Measurement of Meaning*. Urbana: University of Illinois Press.

Pacey, A. (1999). *Meaning in Technology*. Cambridge: MIT Press.

Parr, J. (2010). *Sensing Changes: Technologies, Environments, and the Everyday, 1953–2003*. Vancouver: UBC Press.

Parr, J. (2015). The Senses and the History of Technology. *Technikgeschichte*, *82*(1), 11–25.

Pasveer, B. (1989). Knowledge of Shadows: The Introduction of X-Ray Images in Medicine. *Sociology of Health & Illness*, *11*(4), 360–381.

Pasveer, B. (2006). Representing or Mediating: A History and Philosophy of X-Ray Images in Medicine. In L. Pauwels (Ed.), *Visual Cultures of Science: Rethinking Representational Practices in Knowledge Building and Science Communication* (pp. 41–62). Lebanon, NH: University Press of New England.

Pellico, L., Duffy, T., Fennie, K., & Swan, K. (2012). Looking Is Not Seeing and Listening Is Not Hearing: Effect of an Intervention to Enhance Auditory Skills of Graduate-Entry Nursing Students. *Nursing Education Perspectives*, *33*(4), 234–239.

Pickstone, J. V. (2000). *Ways of Knowing: A New History of Science, Technology and Medicine*. Manchester: Manchester University Press.

Pickstone, J. V. (2007). Working Knowledges Before and After Circa 1800: Practices and Disciplines in the History of Science, Technology, and Medicine. *ISIS*, *98*(3), 451–489.

Pinch, T., & Bijsterveld, K. (2012a). New Keys to the World of Sound. In T. Oxford (Ed.), *Handbook of Sound Studies* (pp. 3–35). Oxford: Oxford University Press.

Pinch, T., & Bijsterveld, K. (Eds.). (2012b). *The Oxford Handbook of Sound Studies*. Oxford: Oxford University Press.

Polanyi, M. (1983/1966). *The Tacit Dimension*. Gloucester, MA: Peter Smith.

Porcello, T. (2004). Speaking of Sound: Language and the Professionalization of Sound Recording Engineers. *Social Studies of Science*, *34*(5), 733–758.

Pot, C. (1933). Klavarskribo: Proeve van een vereenvoudiging van ons notenschrift (Slot). *Maandblad voor Hedendaagsche Muziek*, *2*(4), 129–130.

Potter, R. K., Kopp, G. A., & Green, H. C. (1947). *Visible Speech.* New York: D. van Nostrand Company.

Prescatello, A. R. (1992). *Charles Seeger: A Life in American Music.* Pittsburgh and London: University of Pittsburgh Press.

Rawlinson, A. (1923). *The Defence of London 1915–1918* (2nd ed.). London: Andrew Melrose.

Reichert, S., Gass, R., Brandt, C., & Andrès, E. (2008). Analysis of Respiratory Sounds: State of the Art. *Clinical Medical Insights: Circulatory, Respiratory and Pulmonary Medicine, 2,* 45–58.

Rice, T. (2008). "Beautiful Murmurs": Stethoscopic Listening and Acoustic Objectification. *Senses & Society, 3*(3), 293–306.

Rice, T. (2010). "The Hallmark of a Doctor": The Stethoscope and the Making of Medical Identity. *Journal of Material Culture, 15*(3), 287–301.

Rice, T. (2013). *Hearing and the Hospital: Sound, Listening, Knowledge and Experience.* Canon Pyon: Sean Kingston Publishing.

Rice, T. (2016). Sounds Inside: Prison, Prisoners and Acoustical Agency. *Sound Studies, 2*(1), 6–20.

Rice, T., & Coltart, J. (2006). Getting a Sense of Listening: An Anthropological Perspective on Auscultation. *The British Journal of Cardiology, 13*(1), 56–57.

Rieger, S. (2009). *Schall und Rauch: Eine Mediengeschichte der Kurve.* Frankfurt am Main: Suhrkamp Verlag.

Roberts, L. (1995). The Death of the Sensuous Chemist: The "New" Chemistry and the Transformation of Sensuous Technology. *Studies in the History and Philosophy of Science, 26*(4), 503–529.

Roosth, S. (2009). Screaming Yeast: Sonocytology, Cytoplasmic Milieus, and Cellular Subjectivities. *Critical Inquiry, 35*(2), 332–350.

Ross, C. D. (2004). Sight, Sound, and Tactics in the American Civil War. In M. Smith (Ed.), *Hearing History: A Reader* (pp. 267–278). Athens: University of GeorgiaPress.

Russolo, L. (1986/1916). *The Art of Noises.* New York: Pendragon Press.

Samuels, D. W., Meintjes, L., Ochoa, A. M., & Porcello, T. (2010). Soundscapes: Toward a Sounded Anthropology. *Annual Review of Anthropology, 39,* 329–345.

Scarth, R. N. (1999). *Echoes from the Sky: A Story of Acoustic Defence.* Hythe: Hythe Civic Society.

Schafer, R. Murray. (1967). *Ear Cleaning: Notes for an Experimental Music Course.* Toronto: Berandol Music Limited.

Schafer, R. M. (1994/1977). *The Soundscape: Our Sonic Environment and the Tuning of the World.* Rochester, VT: Destiny Books.

Schirrmacher, A. (2016). Sounds and Repercussions of War: Mobilization, Invention and Conversion of First World War Science in Britain. *France and Germany. History and Technology, 32*(3), 269–292.

Schmandt, B., Jacobsen, S. D., Becker, T. W., Liu, Z., & Dueker, K. G. (2014). Dehydration Melting at the Top of the Lower Mantle. *Science, 344*(6189), 1265–1268.

Schneider, M. (2008). Database Concept for Medical Auditory Alarms. In *Proceedings of the 14th International Conference on Auditory Display, Paris, France, June 24–27* (pp. 1–8).

Schrimshaw, W. (2015). Exit Immersion. *Sound Studies: An Interdisciplinary Journal, 1*(1), 155–170.

Schwartz, H. (2011). *Making Noise: From Babel to the Big Bang & Beyond.* New York: Zone Books.

Schwartz, H. (2012). Inner and Outer Sancta: Earplugs and Hospitals. In T. Pinch & K. Bijsterveld (Eds.), *The Oxford Handbook of Sound Studies* (pp. 273–297). Oxford: Oxford University Press.

Sennett, R. (2008). *The Craftsman.* New Haven, CT: Yale University Press.

Shapin, S., & Schaffer, S. (1985). *Leviathan and the Air-Pump: Hobbes, Boyle and the Experimental Life.* Princeton, NJ: Princeton University Press.

Slabbekoorn, H., & Peet, M. (2003). Ecology: Birds Sing at a Higher Pitch in Urban Noise. *Nature, 424*(6946), 267.

Snook, C. (1925). Automobile-Noise Measurement. *The Journal of the Society of Automobile Engineers, 17*(1), 115–124.

Solomon, L. N. (1954). *A Factorial Study of the Meaning of Complex Auditory Stimuli (Passive Sonar Sounds).* Urbana: University of Illinois.

Sovijärvi, A. R. A., Dalmasso, F., Vanderschoot, J., Malmberg, L. P., Righini, G., & Stoneman, S. A. T. (2000). Definition of Terms for Applications of Respiratory Sounds. *European Respiratory Review, 10*(77), 597–610.

Stafford, B. M. (1994). *Artful Science: Enlightenment Entertainment and the Eclipse of Visual Education.* Cambridge: MIT Press.

Stangl, B. (2000). *Ethnologie im Ohr: die Wirkungsgeschichte des Phonographen.* Vienna: WUV Universitätsverlag.

Sterne, J. (2003). *The Audible Past: Cultural Origins of Sound Reproduction.* Durham: Duke University Press.

Sterne, J. (2009). The Preservation Paradox in Digital Audio. In K. Bijsterveld & J. van Dijck (Eds.), *Sound Souvenirs: Audio Technologies, Memory and Cultural Practices* (pp. 55–65). Amsterdam: Amsterdam University Press.

Stockfelt, O. (1997). Adequate Modes of Listening. In D. Schwarz, A. Kassabian, & L. Siegel (Eds.), *Keeping Score: Music, Disciplinarity, Culture* (pp. 129–146). Charlottesville: University Press of Virginia.

Stockmann, D. (1979). Die Transkription in der Musikethnologie: Geschichte, Probleme, Methoden. *Acta Musicologica, 51*, Fasc. 2, 204–245.

Subotnik, R. R. (1991). *Developing Variations: Style and Ideology in Western Music.* Minneapolis: University of Minnesota Press.

Subotnik, R. R. (1996). Toward a Deconstruction of Structural Listening: A Critique of Schoenberg, Adorno, and Stravinsky. *Deconstructive Variations: Music and Reason in Western Society* (pp. 148–176). Minneapolis: University of Minnesota Press.

Suchman, L. (1987). *Plans and Situated Actions: The Problem of Human-Machine Communication.* New York: Cambridge University Press.

Summers, E. (1916). Notation of Bird Songs and Notes. *The Auk, 33*(1), 78–80.

Supper, A. (2012a). The Search for the "Killer Application": Drawing the Boundaries Around the Sonification of Scientific Data. In T. Pinch & K. Bijsterveld (Eds.), *The Oxford Handbook of Sound Studies* (pp. 127–150). Oxford: Oxford University Press.

Supper, A. (2012b). *Lobbying for the Ear: The Public Fascination with and Academic Legitimacy of the Sonification of Scientific Data* (Ph.D. thesis, Maastricht University).

Supper, A. (2012c). "Trained Ears" and "Correlation Coefficients": A Social Science Perspective on Sonification. In M. A. Nees, B. W. Walker, & J. Freeman (Eds.) *Proceeedings of the 18th International Conference on Auditory Display, Atlanta, GA, USA, June 18–21* (pp. 29–35).

Supper, A. (2014). Sublime Frequencies: The Construction of Sublime Listening Experiences in the Sonification of Scientific Data. *Social Studies of Science, 44*(1), 34–58.

Supper, A. (2015a). Sound Information: Sonification in the Age of Complex Data and Digital Audio. *Information & Culture: A Journal of History, 50*(4), 441–464.

Supper, A. (2015b). Data Karaoke: Sensory and Bodily Skills in Conference Presentations. *Science as Culture, 24*(4), 436–457.

Supper, A. (2016). Lobbying for the Ear, Listening with the Whole Body: The (Anti-)Visual Culture of Sonification. *Sound Studies: An Interdisciplinary Journal, 2*(1), 69–80.

Supper, A., & Bijsterveld, K. (2015). Sounds Convincing: Modes of Listening and Sonic Skills in Knowledge Making. *Interdisciplinary Science Reviews, 40*(2), 124–144.

Takeuchi, A., Hirose, M., Shinbo, T., Imai, M., Mamorita, N., & Ikeda, N. (2006). Development of an Alarm Sound Database and Simulator. *Journal of Clinical Monitoring and Computing, 20*(5), 317–327.

Te Hennepe, M. (2007). *Depicting Skin: Visual Culture in Nineteenth-Century Medicine* (Ph.D. thesis, Maastricht University).

Thompson, E. (2002). *The Soundscape of Modernity: Architectural Acoustics 1900–1933.* Cambridge: MIT Press.

Thorpe, W. H. (1958). The Learning of Song Patterns by Birds, with Especial Reference to the Songs of the Chaffinch 'Fringilla Coelebs'. *The Ibis, 100*(4), 535–570.

Tyler, S. A. (1984). The Vision Quest in the West, or What the Mind's Eye Sees. *Journal of Anthropological Research, 40*(1), 23–40.

Tufte, E. R. (1997). *Visual Explanations: Images and Quantities, Evidence and Narrative.* Cheshire, CT: Graphics Press.

Truax, B. (2001/1984). *Acoustic Communication* (2nd ed.). Westport: Greenwood.

Van der Voort, A. W. M., & Aarts, R. M. (2009). Development of Dutch Sound Locators to Detect Airplanes (1927–1940). In *Proceedings NAG/DAGA, Rotterdam, The Netherlands, March 23–26* (pp. 250–253).

Van Drie, M. (2013). Training the Auscultative Ear: Medical Textbooks and Teaching Tapes (1950–2010). *The Senses and Society, 8*(2), 165–191.

Vertesi, J. (2014). Drawing as: Distinctions and Disambiguation in Digital Images of Mars. In C. Coopmans, J. Vertesi, M. Lynch, & S. Woolgar (Eds.), *Representation in Scientific Practice Revisited* (pp. 15–35). Cambridge: MIT Press.

Vickers, P. (2011). Sonification for Process Monitoring. In T. Hermann, A. Hunt, & J. G. Neuhoff (Eds.), *The Sonification Handbook* (pp. 455–491). Berlin: Logos Verlag.

Vickers, P. (2012). Ways of Listening and Modes of Being: Electroacoustic Auditory Display. *Journal of Sonic Studies, 2*, 1. Available at http://journal.sonicstudies.org/vol02/nr01/a04. Last accessed August 18, 2017.

Volmar, A. (2012). *Klang als Medium wissenschaftlicher Erkenntnis: Eine Geschichte der auditiven Kultur der Naturwissenschaften seit 1800* (Ph.D. thesis, Universität Siegen).

Volmar, A. (2013). Listening to the Cold War: The Nuclear Test Ban Negotiations, Seismology, and Pyschoacoustics, 1958–1963. In A. Hui, J. Kursell & M. Jackson (Eds.), *Music, Sound and the Laboratory from 1750–1980. Osiris, 28*, 80–102.

Volmar, A. (2014). In Storms of Steel: The Soundscape of World War I and Its Impact on Auditory Media Culture During the Weimar Period. In D. Morat (Ed.), *Sounds of Modern History: Auditory Cultures in 19th- and 20th-Century Europe* (pp. 227–255). New York, NY: Berghahn.

Volmar, A. (2015). Ein "Trommelfeuer von akustischen Signalen": Zur auditiven Produktion von Wissen in der Geschichte der Strahlenmessung. *Technikgeschichte, 82*(1), 27–46.

Walker, B. N., & Nees, M. A. (2011). Theory of Sonification. In T. Hermann, A. Hunt, & J. G. Neuhoff (Eds.), *The Sonification Handbook* (pp. 9–39). Berlin: Logos Verlag.

Wellek, A. (1932). Die Entwicklung unserer Notenschrift aus dem Tönesehen. *Acta Musicologica, 4*(3), 114–123.

Whittaker, W. G. (1924). The Claims of Tonic Solfa I. *Music & Letters, 5*(4), 313–321.

Williams, S. M. (1994). Perceptual Principles in Sound Grouping. In G. Kramer (Ed.), *Auditory Display: Sonification, Audification, and Auditory Interfaces* (pp. 95–125). Reading: Addison-Wesley Publishing Company.

Willaert, T. (2016). *De fonograaf en de grammofoon in de Nederlandstalige literatuur 1877–1935: Een media-archeologisch onderzoek* (Ph.D. thesis, Leuven University).

Woolf, D. R. (2004). Hearing Renaissance England. In M. M. Smith (Ed.), *Hearing History: A Reader* (pp. 112–135). Athens: University of Georgia Press.

Woolf, C. R., & Rosenberg, A. (1964). Objective Assessment of Cough Suppressants Under Clinical Conditions Using a Tape Recorder System. *Thorax, 19*(2), 125–130.

Worrall, D. (2009). *Sonification and Information: Concepts, Instruments and Techniques* (Ph.D. thesis, University of Canberra).

Zwikker, C. (1934). De oorzaken van het geluid bij automobielen. In V. Anonymous (Ed.), *van het 'Anti-lawaai Congres', georganiseerd te Delft, op 8 november 1934 door de Koninklijke Nederlandsche Automobile Club in samenwerking met de Geluidstichting* (pp. 70–77). Delft: KNAC/Geluidstichting.

Name Index

Subject Index

Printed in the United States
by Bookmasters

Printed in the United States
By Bookmasters